5th Course on Materials for Energy and Sustainability (MRS and E-MRS) and 3rd Course of the EPS SIF International School on Energy (LNES 2016)

EPJ Web of Conferences Volume 148 (2017)

Erice, Italy
13 - 19 July 2016

Editors:

Antonio Terrasi
David Cahen
Luisa Cifarelli
David S. Ginley
Abdelilah Slaoui
Friedrich Wagner

ISBN: 978-1-5108-4514-5

Printed from e-media with permission by:

Curran Associates, Inc.
57 Morehouse Lane
Red Hook, NY 12571

Some format issues inherent in the e-media version may also appear in this print version.

Printed by Curran Associates, Inc. (2017)

For additional information, please contact EDP Sciences – Web of Conferences at the address below.

EDP Sciences – Web of Conferences
17, Avenue du Hoggar
Parc d'Activité de Courtabœuf
BP 112
F-91944 Les Ulis Cedex A
France

Phone: +33 (0) 1 69 18 75 75
Fax: +33 (0) 1 69 28 84 91

contact-edps@webofconferences.org

Additional copies of this publication are available from:

Curran Associates, Inc.
57 Morehouse Lane
Red Hook, NY 12571 USA
Phone: 845-758-0400
Fax: 845-758-2633
Email: curran@proceedings.com
Web: www.proceedings.com

TABLE OF CONTENTS

Preface

Since 2010, the Ettore Majorana Foundation and Centre for Scientific Culture (Erice, Italy) has been holding a biannual school on renewable energies and sustainability co-organized by MRS and E-MRS. The Ettore Majorana Foundation has been sponsoring world-class schools for more than 50 years. The aim of the school is to present the state of the art and future perspectives for materials applied to the generation and storage of renewable and sustainable energy. This year's *International School of Solid State Physics* was held on 13–19 July with the fifth course on "Materials for Energy and Sustainability" (MRS and E-MRS). This time, the European Physical Society (EPS) and the Italian Physical Society (SIF) merged the third course of the *Joint EPS-SIF International School on Energy* (a biannual school usually taking place in Varenna, Italy) with this school. Also the Eurosunmed Project joined the organization of the school, all adding valuable contributions to this event.

Seventy-five PhD students and young researchers from 14 different countries experienced seven days of "full immersion" in the field of energy and sustainability. They attended lectures covering some of the most important and strategic topics in the field of generation, storage, and consumption of energy, which aimed at putting these topics in perspective in terms of sustainability, with an emphasis on lifecycles. In addition, the students were given projects, which they worked on in small groups and reported on at the end of the school.

Thanks to distinguished speakers from France, Germany, Israel, Italy, The Netherlands, Switzerland, and the United States, the school gave a wide overview of energy challenges, including technologies, critical materials, pollution, safety, and economy. Teaching was accomplished through lectures and close interactions with students during the breaks, lunches, dinners, and poster sessions. Lectures focused on global climate, CO_2 sequestration, water-energy nexus, nuclear energy, solar power, both thermal and photovoltaic, as well as photosynthesis and biofuels, geothermal and wind energy, various forms of energy storage, thermoelectricity, the future of the electrical grid, and how to view mobility and buildings from the point of view of energy and sustainability.

Students were continuously stimulated to interact (for example, by having different neighbors during the lectures each day) and to solve problems and exercises. Their main target was to gather in teams of five to seven people and critically assess papers from *MRS Energy & Sustainability —A Review Journal*, which publishes reviews on key topics in materials research and development as they relate to energy and sustainability.

At the end of the school, each team presented its analysis of "their" paper, having done fact- and logic-checking (*i.e.*, serving as critical readers). An important aim of the school was to provide the students with the background needed to have a critical, not passive, attitude when involved in their field of research while using their familiarity with the scientific method.

The school is a great opportunity for students and postdoctoral fellows from around the world to meet with and learn from their peers and established experts in their area of study.

Most of the speakers (14 over 21) were also able to provide a manuscript based on their lectures as a further support for present and future students. These *Lecture Notes* are here reported as an on-line updated source of information in the amazing and exciting field of renewable and sustainable energies.

A. Terrasi, D. Cahen, L. Cifarelli,
D. Ginley, A. Slaoui and F. Wagner

edp sciences EPJ Web of Conferences

Statement of Peer review

In submitting conference proceedings to EPJ Web of Conferences, I certify to the Publisher that I adhere to the **Policy on Publishing Integrity** of the journal in order to safeguard good scientific practice in publishing.

1. All articles have been subjected to peer review administered by the proceedings editors.
2. Reviews have been conducted by expert referees, who have been requested to provide unbiased and constructive comments aimed, whenever possible, at improving the work.
3. Proceedings editors have taken all reasonable steps to ensure the quality of the materials they publish and their decision to accept or reject a paper for publication has been based only on the merits of the work and the relevance to the journal.

Title, date and place of the conference

5th course of the MRS-EMRS *"Materials for Energy and Sustainability"* and 3rd course of the *"EPS-SIF International School on Energy"*.

13-19 July 2016, Erice (Italy)

Proceedings editor(s):

Antonio Terrasi, David Cahen, Luisa Cifarelli, David S. Ginley, Abdelilah Slaoui, Friedrich Wagner

Date and editor's signature

June 14th 2017

EPJ Web of Conferences **148**, 00002 (2017) DOI: 10.1051/epjconf/201714800002
LNES 2016

The global carbon nation: Status of CO_2 capture, storage and utilization

Elizabeth A. Kocs

University of Illinois at Chicago - IL, USA and
Energy Initiative - Erice (TP), Italy

Summary. — As the world transitions toward cleaner and more sustainable energy generation, Carbon Capture and Sequestration/Storage (CCS) plays an essential role in the portfolio of technologies to help reduce global greenhouse gas (GHG) emissions. The projected increase in population size and its resulting increase in global energy consumption, for both transportation and the electricity grid —the largest emitters of greenhouse gases, will continue to add to current CO_2 emissions levels during this transition. Since eighty percent of today's global energy continues to be generated by fossil fuels, a shift to low-carbon energy sources will take many decades. In recent years, shifting to renewables and increasing energy efficiencies have taken more importance than deploying CCS. Together, this triad —renewables, energy efficiency, and CCS— represent a strong paradigm for achieving a carbon-free world. Additionally, the need to accelerate CCS in developing economies like China and India are of increasing concern since migration to renewables is unlikely to occur quickly in those countries. CCS of stationary sources, accounting for only 20% reduction in emissions, as well as increasing efficiency in current systems are needed for major reductions in emissions. A rising urgency for fifty to eighty percent reduction of CO_2 emissions by 2050 and one hundred percent reduction by 2100 makes CCS all that more critical in the transition to a cleaner-energy future globally.

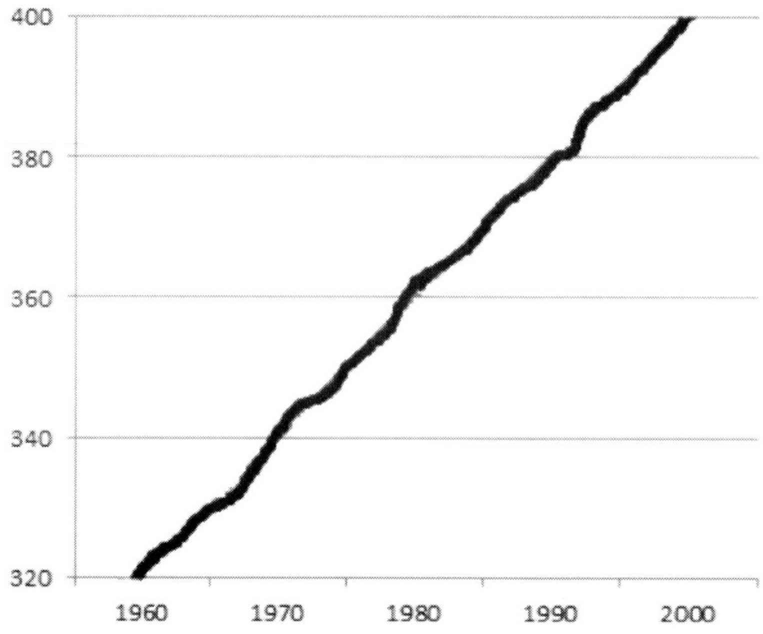

Fig. 1. – 30% rise in atmospheric CO_2 levels in 50 years (1960 to 2013.) (Source: Global CCS Institute 2015 Report & Mauna Loa Observatory).

1. – Background

Increasing use of fossil fuels, such as coal, oil and gas, since the mid-18th century, marked by the beginning of the Industrial Revolution, has led to a steady increase in carbon dioxide (CO_2) in the Earth's atmosphere. The last fifty years have demonstrated an intense increase that has put the rate of CO_2 permanently over 400 parts per minute (ppm) [1]. (See fig. 1.)

The world continues to be reliant on fossil fuels for transportation and electricity generation. In fact, the impact of a growing global population predicted to rise to 9 billion by 2050 is expected to increase world energy demand by fifty percent [2]. Reducing the global greenhouse (GHG) emissions from these sources cannot be achieved solely through increasing energy efficiency and deploying more renewable energy production [3]. Based on United States Environmental Protection Agency (EPA) estimates, a passenger car emits a little over 5000 kg of CO_2 a year (5 metric tonnes). Twenty-two million tonnes of stored CO_2, therefore, is equivalent to mitigating 4.2 million cars. (See fig. 2.)

Global CO_2 emissions from transportation and the electricity sector can be achieved in a number of ways. Improving energy efficiency by requiring people and industries to waste less energy, utilizing methods and techniques to enhance current systems, and changing behavior to use energy more efficiently is one area that can and has seen considerable impact on both energy consumption and resulting emissions reductions. Increasing the

DOI: 10.1051/epjconf/201714800002

Fig. 2. – CCS can be an efficient means of large emissions reductions. (Source: EPA).

deployment of renewable energy sources, such as wind, solar, wave and geothermal is another area where the transition to a cleaner-energy future has and will continue to have influence on emissions. Lastly, emissions reduction through capture and storage and/or reuse can help by preventing additional emissions from entering the atmosphere.

Over sixty percent of energy use and GHG emissions are known to be from the electricity sector and transportation [4]. (See fig. 3.) There is a direct relationship between energy use and CO_2 emissions. Greater use of fossil fuel energy production

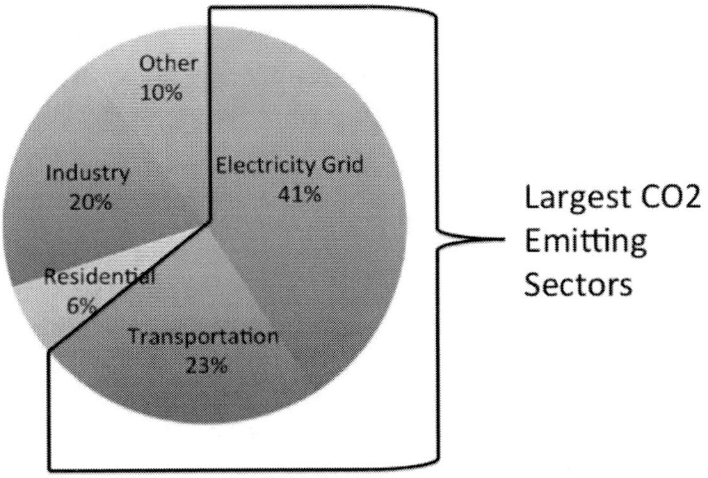

Fig. 3. – CO_2 emission by sector (Source: IEA).

leads to higher CO_2 emissions. Gasoline cars replaced by electric or biofuels ones and the transition from fossil fuel power plants to renewable electricity generation will change the amount of GHG emitted by these sources. CCS allows the continued use of fossil fuels during the decades it will take for the entire world to transition to non-carbon-dioxide-emitting energy sources. As noted earlier, the world still relies on fossil fuels today and it will remain the main source of energy generation for decades to come [5]. Fossil fuels (coal, oil, gas) represent eighty percent of the global energy mix, while renewables only account for thirteen percent, and are only projected to contribute to thirty percent of the global energy mix by 2030 [6].

2. – Why is CO_2 capture and storage important?

Technologies for carbon dioxide capture and storage have the potential to reduce carbon dioxide emissions from energy generated by fossil fuels. As noted earlier, CCS is not sufficient on its own to achieve the fifty to eighty percent reduction in emissions needed by 2050 and the one hundred percent needed by 2100. To achieve this type of emissions reduction, CCS must work in conjunction with a transition to renewable energy generation and increased energy efficiency. Since more than eighty percent of today's energy comes from fossil fuels and the transition to low or no carbon energy sources is time consuming, challenging and costly, CCS is necessary to achieve large and rapid carbon dioxide emissions reductions.

In the United States alone, more than 40% of CO_2 emissions are from power generation for the electricity grid. CCS can reduce emissions from electric power generation by eighty to ninety percent, a dramatic decrease that can have broad impacts across the electricity sector. Industrial sectors, such as natural gas production, cement, steel, and chemicals, can also benefit greatly from implementation of CCS by reducing emissions from industrial processes [7]. These industries typically generate CO_2 during production processes, in addition to using it as fuel, and have no alternatives for reducing emissions. (See fig. 4.) Implementing CCS technologies in the industrial sector will allow for these industries to continue to be viable and bring economic benefits [8]. Additionally, countries that have the geographic capacity to provide storage for captured CO_2 have the potential to benefit from an emerging and needed market.

The growing and continued demand for energy makes CCS a needed and desirable solution to global CO_2 emissions, given projections that fossil fuels will continue to be needed in the transition to renewables over the coming decades [9]. CCS can reduce emissions from many sources, and is applicable to sixty percent (15 billion tonnes) of CO_2 emissions per year coming from stationary sources such as power plants and industry. (See fig. 5.) Over seven thousand four hundred sources greater than 0.1 Mt/yr. currently exist globally. CCS continues to expand worldwide with approximately fourteen megatonnes per year coming from operating industrial scale projects. Thirteen megatonnes per year are currently under construction to expand the industrial scale projects [10].

Global energy-related CO_2 emissions coming from industrial applications are currently at about twenty-five percent of the total and are projected to grow to more than 50% by

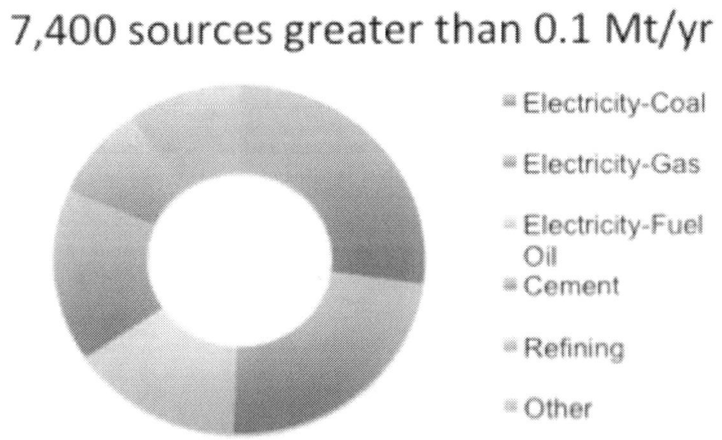

Fig. 4. – 7400 sources greater than 0.1 Mt/yr (Source: IEA).

2050, if emissions are not addressed as a more critical issue [11]. Successful demonstration of CCS in industrial sectors is and will continue to be vital to future emission reduction efforts. Currently, around 40 million tonnes per year (Mtpa) —similar to the total annual CO_2 emissions of Denmark or Switzerland— are being captured by twenty-two projects around the globe.

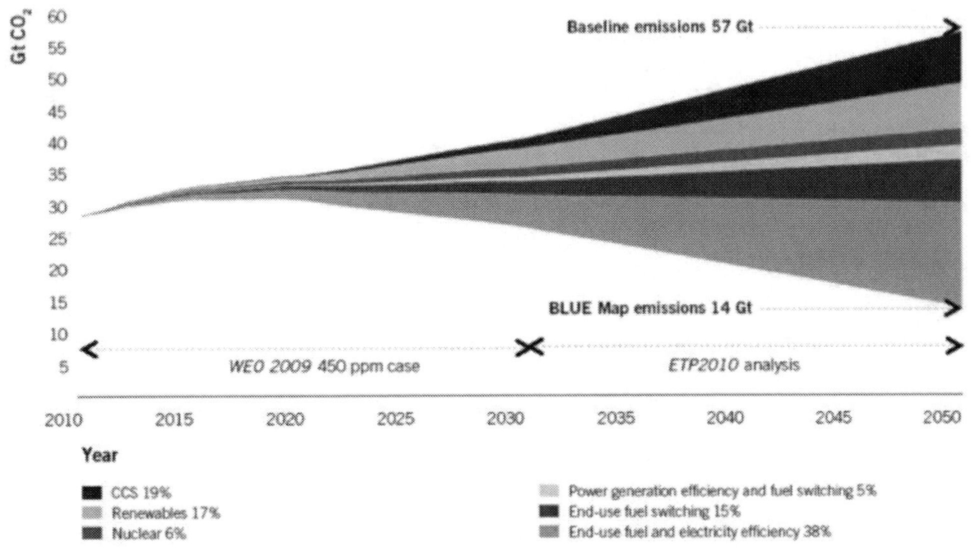

Fig. 5. – CCS is expected to contribute about 20% to CO_2 reductions. (Source: IEA).

3. – CCS is expected to contribute to the needed CO_2 emissions reduction

De-carbonizing the power and transportation sectors by controlling emissions as well as reducing the demand for fossil fuel energy and supplying low- or no-carbon energy continues to be the key to a clean energy future. Developing countries such as China and India have been the largest contributors to increasing CO_2 emissions due to the use of coal, the dirtiest and most easily accessible fossil fuel worldwide. Similar trends in other developing countries could potentially double or triple CO_2 emissions by mid-century, making CCS even more critical in working toward reversing or mitigating the existing GHG emission problem [12]. CCS is not without controversy, which will be discussed later, where questions about its stability to remain stored is an actuality over longer time periods. CCS technology exists and can continue to be perfected. What remains to be seen is whether CCS finds itself equal to other strategies, such as increased deployment of renewables and enhanced energy efficiency, and how quickly and broadly it can be implemented. The CO_2 in the atmosphere is not going anywhere, and in fact, will only continue to rise unless CCS is deployed where it counts the most —the power and industrial sectors.

4. – CCS is an efficient means of large emissions reductions

Understanding the scale and scope of CCS is important for realizing the value it adds. CCS that can capture ninety percent of CO_2, *i.e.*, one 1000 MW coal-fired power plant which emits approximately 6.5 MT of CO_2 per year is equal to an increased car efficiency from 25 to 50 mpg of 2.8 million cars, which is about ten percent of California's personal vehicle fleet (see fig. 2). At this scale, CCS has greater significance for increasing reductions through both the number of ways and locations for greatest impact. In looking at new power plants (typically natural gas) *versus* old power plants (many of which are coal-fired plants), newer natural gas plants can utilize efficient generation technology to achieve CO_2 reduction standards, whereas new coal plants cannot and must use CCS to reach CO_2 reduction standards [13].

Currently, fifteen active commercial-scale industrial CCS projects, ten of them in the United States, are currently deployed worldwide including the Boundary Dam Power Station in Saskatchewan, Canada in operation since late 2014, and other coal-fired commercial power plants, such as NRG Energy's Petra Nova project in Texas opening in 2017. Globally, another fifty commercial-scale CCS projects will come online in the electricity and industrial sectors in the coming decade.

For the United States, and for the world, all options need to be considered in the race against GHG emissions. All efforts, expanding renewables, enhancing energy efficiency and conservation, and even creating a carbon trade market, are part of the solution along with CCS, and will need to be implemented simultaneously to have the greatest impact. (See fig. 6.) CCS can serve as the bridge to a low-carbon energy future and economy by reducing carbon dioxide emissions from power generation while expanding renewables and increasing energy efficiency [14].

 DOI: 10.1051/epjconf/201714800002

Fig. 6. – Quote from IEA.

5. – The opportunity

CCS alone can provide up to 20% of the CO_2 emission reductions we need to make by 2050, and is the only way to greatly reduce carbon emissions from electricity generation while working toward expansion of renewable energy capacity and increasing energy efficiency and conservation. (See fig. 7.) Many question the necessity of CCS, given the success of energy efficiency and the investments in expanding renewable energy. Renewable energy sources currently account for minimal electricity generation (just four percent for the United States) and would be highly unlikely to be deployed at a rate that could meet large portions of energy demands within the next decade. In fact, such large-scale transitions to renewables are projected to take place across multiple decades as investments increase in renewables and older, less efficient, and dirtier power generation

Fig. 7. – Role and Opportunity for CCS in clean energy future (Image after Zero Emissions Platform (ZEP), Source: `http://www.zeroemissionsplatform.eu/`).

 DOI: 10.1051/epjconf/201714800002

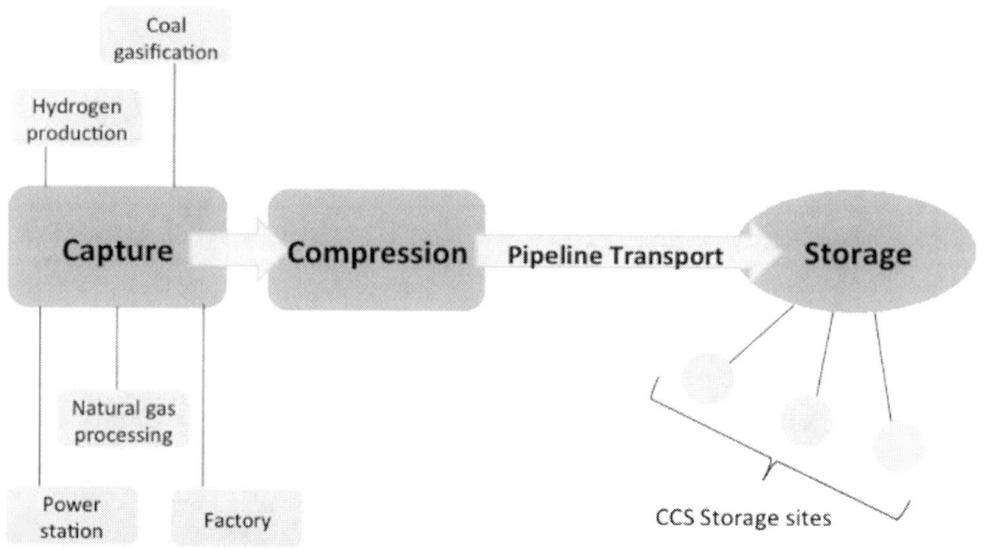

Fig. 8. – CCS Network (Source: WRI).

systems become retired [14]. The time gap for this transition as well as the existing global reliance on coal allows CCS to step in to provide needed CO_2 reductions while continuing to meet future energy demands.

6. – How does CCS work?

CO_2 Capture and Storage involves 4 steps. (See fig. 8.) CCS is a system of technologies that captures carbon dioxide from power plants and industrial sources, compresses it, transports it typically via pipeline to appropriate locations, and injects the CO_2 into geological formations deep below the subsurface of the Earth for long-term sequestration [14]. At least ninety percent of emissions from fixed emitters can be captured and stored safely and permanently using natural trapping mechanisms. Transporting CO_2 is not new. It has been around for decades.

Three technologies exist to capture CO_2 —pre-combustion, where CO_2 is captured before fuel is burned; oxy-fuel, where CO_2 is captured during fuel combustion; and post-combustion, where CO_2 is captured after fuel has been burned (this technology can be retrofitted to existing power and industrial plants). (See table I.)

Cost and performance of today's CCS technology has its limits. Firstly, there is immediately an energy penalty of ten to thirty percent, meaning that energy is lost to the CCS process, reducing efficiencies for the overall system. Cost is the second largest limitation to implementing CCS. Typically the cost of integrating CCS into power plants ranges from \$60 to \$110 per tonnes of CO_2 for an "X" number of power plants, and yet the first plant will always cost more, anywhere from \$150 to \$250 per tonnes of CO_2.

DOI: 10.1051/epjconf/201714800002

TABLE I. – *Comparison of capture options (Source: Global CCS Institute 2015 Report).*

Technology	Advantages	Challanges
Post-combustion	• Mature technology • Standard retrofit	• High energy penalty (20–30%) • High cost for capture
Oxygen-Combustion	• Avoid complex post-combustion separation • Potentially higher generation efficiencies	• Oxygen separation • Repowering
Pre-Combustion(IGCC)	• Lower costs than post-combustion • Lower energy penalties (10–15%) • H2 production	• Complex chemical process • Repowering • Large capital investment

This additional cost for each power plant can increases the cost of electricity generation from fifty to one hundred percent. Thirdly, reliability remains a question for both the system's ability to capture the CO_2 but also for the ability of the system to store the CO_2 for indefinite lengths of time without impacting any other Earth system while being stored. (See table II for technological advantages and challenges of advanced materials and processes for CO_2 capture.) Further research and development is needed not only to advance existing CCS technologies but also to find new and better ways to utilize the CO_2.

Other risks and limitations to CCS continue to raise more questions than provide the necessary answers. Many are concerned about the capacity to sequester all this carbon dioxide, and whether it will really remain underground indefinitely. Geological formations have stored CO_2 oil and gas for millions of years and would be highly likely to sequester CO_2 as well indefinitely with the proper site selection and post-storage management. Many existing CCS sites have stored CO_2 for well over a decade without any leakage [14]. This does not mean CCS is without risk. In fact, the risks are comparable to other industries that utilize underground storage, such as natural gas. Like all risks, they can be managed and more importantly weighed against the alternative, emitting the CO_2 into the atmosphere. This is where best practices, regulatory oversight and development of guidelines for siting, monitoring, and caring for CO_2 storage sites over the long-term become critical to the success of CCS.

The challenges for CCS are three-fold —technical, regulatory and economic. CCS will certainly be more accepted by society if it can be shown to be safe and effective, as well as economical. The large-scale projects already deployed and the ones under development will play a critical role in determining the future of CCS. Research and technological development, like other large and infrastructure oriented technological advances, will require considerable investment to lower the cost of development and integration, and the

TABLE II. – *Advanced materials and processes for CO_2 capture (Source: De Coninck and Benson, S., 2014).*

Separation Approach	Absorption	Adsorption	Cryogenic	Membranes	Mineralization
Example Materials	Aqueous amine solutions Chilled ammonia Ionic liquids	Zeolites Metal organic frameworks (MOFs) Activated carbon	No specific material requirements	Polymer membranes Inorganic membranes	Magnesium silicates Alkalai-rich waste streams
Advantages	Numerous solvent options Rapid improvements in energy requirements achieved	Potentially lower energy requirements for regeneration	Avoid need for solvents or sorbents Lower energy requirements	Avoid regeneration energy requirements	Co_2 is converted to a solid substrate that can be reused as a building material or diposed of in surface facilities
Technological Challenges	Reducing energy for regeneration Solvent degradation	Adsorption capacity and kinetics	Solid separation and handling	Permeability Selectivity	Rate of reactions Large mass of reactants (*e.g.* source of Mg, Ca)

skills needed to install, maintain and manage the systems. Additionally, a regulatory framework is necessary at various scales, national and international, for responsible parties, rights and liabilities, and for global standards to which each must adhere [9]. Clear policies will provide direction for the value of CCS to the world both from an environmental perspective as well as an economical one.

7. – Prospects for CO_2 storage globally

CO_2 storage resources are currently available to support CCS deployment. (See fig. 9.) They are vast and appear to be greater than the projected capacity requirements for the coming decades [15].

How much CO_2 storage actually exists? (See fig. 10 for potential global CO_2 storage capacity.) There are a few best options for underground CO_2 storage: deep saline aquifers, depleted oil and gas reservoirs (including sub-sea reservoirs) and non-mineable coal seams. When storing CO_2 in underground reservoirs, the bubble of CO_2 is under impermeable rock deeper than 800 meters, which itself is at the very top of a water-filled reservoir rock. With potential capacity for decades or hundreds of years, deep saline aquifers with 1000-10000 Gt or more potential CO_2 storage appears to be the

 DOI: 10.1051/epjconf/201714800002

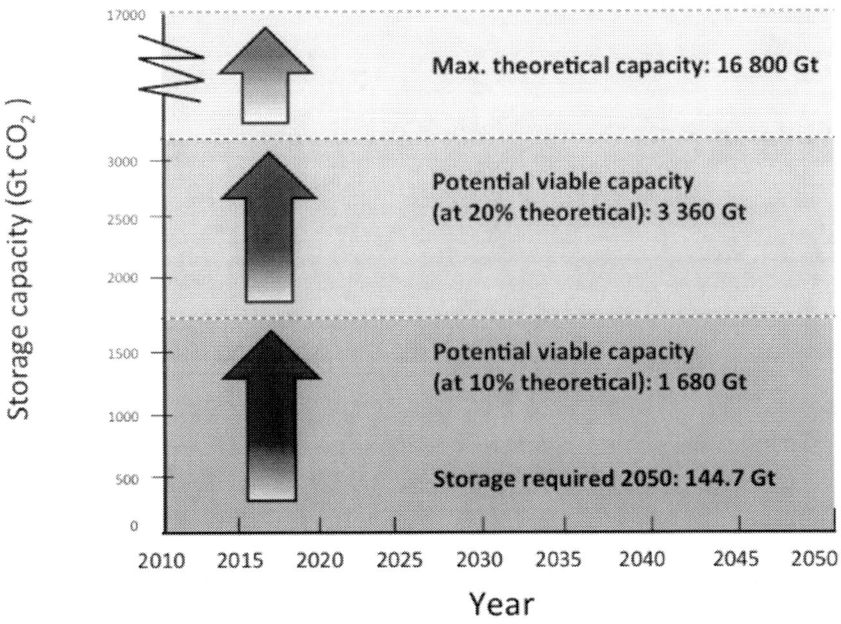

Fig. 9. – World total storage capacity (Source: IEA).

most significant storage source globally. Depleted oil and gas fields can store around 920 Gt. Non-mineable coal seams have considerably smaller potential capacity, however when combined with enhanced oil recovery (EOR), enhanced coal-bed methane recovery

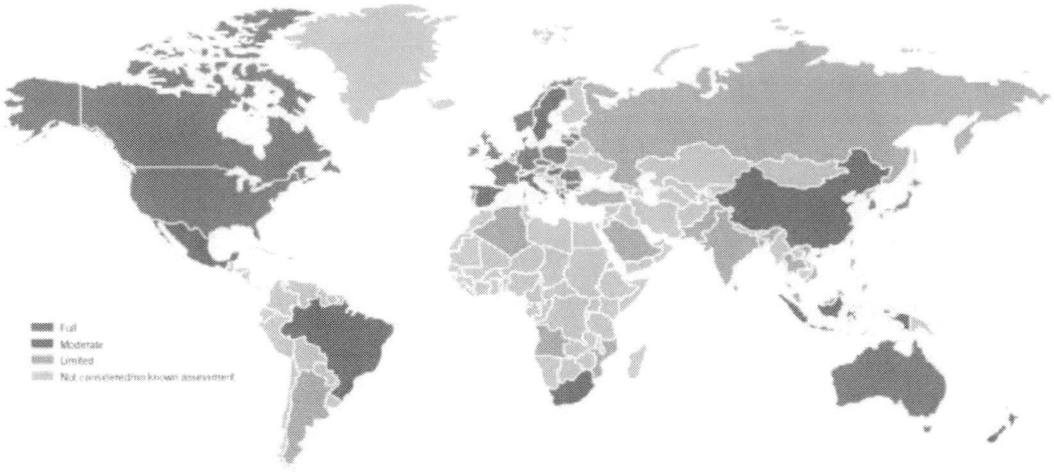

Fig. 10. – Prospects for CO_2 Storage Globally. (Source: Global CCS Institute 2015 Report).

TABLE III. – *Capacity of CO_2 storage options (Source: De Coninck and Benson, S., 2014).*

Reservoir	Gigatons Carbon
Atmosphere	720
Surface Ocean	670
Deep Ocean	36.730
Carbonate Rocks	> 60.000.000
Fossil fuels	4.130

(ECMR), and enhanced gas recovery (EGR) these fields have uncertain but considerably greater potential for CO_2 storage. When used in such combinations, the additional cost of CO_2 transportation and storage can be offset. Some deterrents for utilizing these options include: large distances of storage sites from emissions sources, and high cost of intercontinental transportation of CO_2.

Surface mineralization of CO_2 has the potential to be a permanent and benign form of storage that requires no follow-up or regular monitoring. (See table III for a comparison of capacity of CO_2 storage options.) The most important component of this type of storage is that the capacity far exceeds any current or future emissions. This process as noted earlier is not without its challenges, specifically scientific ones. The process is very slow. The kinetics needed to transform CO_2 to rocks needs to be faster and the right catalysts are yet to be identified. Additionally during the process of mineralization, the surface chemistry still needs to be controlled so as to prevent the CO_2 from releasing or reacting with another airborne particle. For this, ideally a non-reactive coating would need to be developed.

Other options for potential CO_2 storage are: oceanic storage and surface mineralization, however these two are not current viable options since environmental impacts of storing CO_2 in the water column are currently unknown and transforming CO_2 in rocks is still very much in the development stage since mineralization of CO_2 is a lengthy process and catalysts are needed to speed up the process. Besides energy generation and enhanced oil recovery, which are two major contributors to CO_2 emissions, the following industrial sectors should also be considered for CCS: cement, iron and steel, downstream oil, biomass, and other high-purity sources of CO_2 (*e.g.*, liquid natural gas and gas processing) [16].

One topic that has been part of the discussion for CCS is reusing CO_2. Currently the potential for CO_2 reuse is extremely limited, in particular for the chemical industry since the top 100 chemicals produced globally that can utilize CO_2 are only 0.5 Gt/yr, while CO_2 emissions are 35 Gt/yr. Fuels are currently the best option for CO_2 reuse at scale. One major challenge for CO_2 reuse, similar to surface mineralization, is the need for catalysts. Recent research on a novel copper oxide derived catalyst that converts carbon monoxide (CO) to ethanol and acetate at room temperatures shows progress in this area.

DOI: 10.1051/epjconf/201714800002

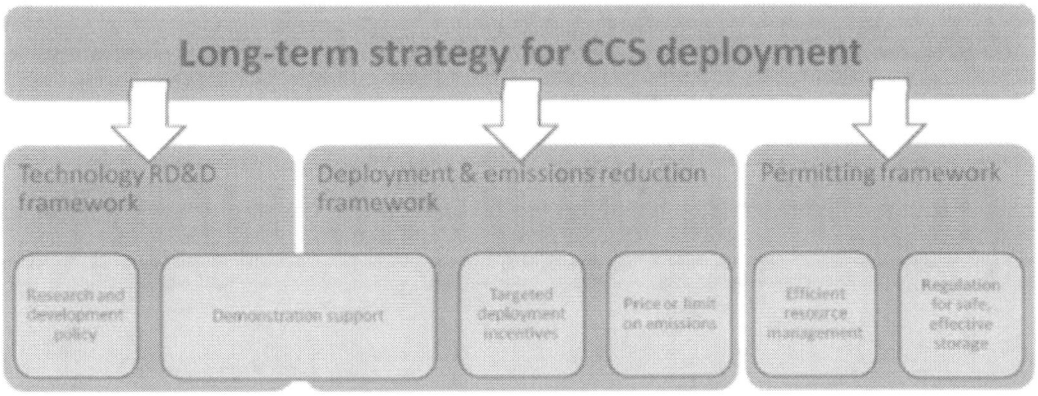

Fig. 11. – Key elements of comprehensive CCS policy framework (Source: IEA).

Key elements for the safety and security of geological storage for CO_2 requires a complete understanding of the following: financial responsibility, regulatory oversight, contingency planning and remediation, monitoring, safe operations, storage engineering, site characterization and selection, and fundamental storage and leakage mechanisms [17]. (See fig. 11 for the diagram of a comprehensive CCS deployment strategy.)

Critical issues for CCS include: gaining practical experience with CCS and power generation, whereby reliability and operating costs are aligned with expectations; lowering the cost of CCS by 50% or more since current technology adds approximately 3–6 cents per kWh of additional cost; increasing societal confidence in the safety and security of CCS, by understanding the impacts of CO_2 retention on groundwater and potential for increased seismic activity; sustaining research and development on the science and technology needed to bring new capture technologies to market and strengthen storage security, site characterization, and monitoring; and introducing a favorable policy and regulatory environment [14].

Since we are not yet in a position to utilize CCS widely, what do we still need for greater deployment? (See table IV for an overview of CCS projects worldwide.) When it comes to policies, it appears relatively straightforward. First and foremost, large- and commercial-scale projects are needed to better inform economic, technical and regulatory decision-making. These projects will provide information for different capture and storage technologies, and better monitoring practices and technologies for storing large volumes of CO_2 for long periods of time. The end result will be lower costs and creating a competitive market for deploying CCS. Secondly, policies that incentivize the use of CCS for power plants would accelerate its deployment as well. Currently, the cost to emit CO_2 is zero, and without both market incentives and a complimentary policy, investments in CCS will be few and far between. Thirdly, regulations are needed to guarantee CCS is a safe process. These regulations have already started to emerge at different levels (*e.g.*, in the U.S. both at national and state level), and work remains to be done on how to

TABLE IV. – *Overview of CCS projects globally (Source: IEA).*

	No. of projects
CO_2 capture demonstration projects	11
CO_2 capture R&D projects	35
Geologic storage projects	26
Geologic storage R&D projects	74
Ocean storage R&D projects	9

address long-term challenges and liabilities as well as rights to property. Lastly, one of the most controversial and challenging of all is public acceptance of CCS. Due to the lack of adequate knowledge, certainty and public dialogue about CCS, very little confidence from the general public exists currently for the successful deployment of CCS. Through addressing these four key areas for policy, CCS can have greater deployment potential and future success [18].

8. – Conclusion

Without any changes to current policies, by 2030 emissions are projected to increase by 63% from today's levels and 90% from 1990 levels. Even with implementation of some global mitigation policies through 2030, the increase can still be up to 40% globally from today's levels [19]. CCS holds promise for reducing emissions and having significant benefits environmentally, economically and for worldwide energy security [12]. There is no one solution to the CO_2 emissions challenges we face globally. A portfolio of solutions is needed (as suggested in fig. 12). Alongside the deployment of more renewables and increasing and enhancing energy efficiency across all sectors, CCS is a critical part of a sustainable energy future, one upon which the climate depends.

Rapid and widespread deployment of CCS across power generation, EOR, and industrial sectors is necessary. Globally, we need to move from the successful, existing small-scale CCS projects in operation today to building at the very least 3 400 commercial-scale projects worldwide by 2050, if CCS is to provide 20% of the CO_2 reductions needed [20].

Fig. 12. – A portfolio of solutions for CO_2 emissions.

On the demand side of the equation, changing behaviors and building infrastructure that uses energy more efficiently could lower the amount of energy needed. "Lifestyle and behavioral changes could reduce energy demand by up to 20 percent in the short term and by up to 50 percent of present levels by mid-century" [12]. Any solutions for CO_2 emissions need to be integral to planning and should seek to create "cities of the future", whereby energy infrastructure improves urban resiliency, sustainability, and regeneration and looks for ways to integrate the potentials of nature with our human demands for energy.

REFERENCES

[1] Petroleum Technology Research Centre and Global Carbone Capture and Storage Institute, in *What happens when CO_2 is stored underground? Q&A from the IEAGHG Weyburn-Midale CO_2 Monitoring and Storage Project* (2014). Retrieved from: `https://hub.globalccsinstitute.com/sites/default/files/publications/151303/co2-stored-underground-ieaghg-weyburn-midale-co2-monitoring-storage-project.pdf`.

[2] United Nations, in *2014 Revision World Urbanization Prospects* (2014). Retrieved from: `https://esa.un.org/unpd/wup/publications/files/wup2014-highlights.Pdf`.

[3] Standeland A., in *Why CO_2 Capture And Storage (CCS) Is An Important Strategy To Reduce Global CO_2 Emissions*, October 9 (2007). Retrieved from: `http://bellona.org/news/ccs/2007-10-why-co2-capture-and-storage-ccs-is-an-important-strategy-to-reduce-global-co2-emissions`.

[4] United States Environmental Protection Agency (EPA), in *Sources of Greenhouse Gas Emissions*, last updated October 6 (2016). Retrieved from: `https://www.epa.gov/ghgemissions/sources-greenhouse-gas-emissions`.

[5] International Energy Agency (IEA), in *Key World Energy Statistics 2009* (2009). Retrieved from: `http://www.observatoire-du-nucleaire.org/IMG/pdf/2009-keyworld-aie.pdf`.

[6] International Energy Agency (IEA), in *World Energy Outlook 2015* (2015). Retrieved from: `https://www.iea.org/Textbase/npsum/WEO2015SUM.pdf`.

[7] United States Environmental Protection Agency (EPA), in *Carbon Dioxide Capture Sequestration*, last updated August 9 (2016). Retrieved from: `https://www3.epa.gov/climatechange/ccs/index.html`.

[8] Carbon Capture and Storage Association (CCS^a), in *Economic Importance* (2012). Retrieved from: `http://www.ccsassociation.org/why-ccs/economic-importance/`.

[9] CO_2 Capture Project (CCP), in *What is CO_2 Capture and Storage?* (2015). Retrieved from: `http://www.co2captureproject.org/what_is_co2_capture_storage.html`.

[10] De Coninck and Benson S., in *Carbon Dioxide Capture and Storage: Issues and Prospects*, *Annu. Rev. Environ. Res.*, **39** (2014) 243. Retrieved from: `http://www.annualreviews.org/doi/abs/10.1146/annurev-environ-032112-095222`.

[11] International Energy Agency (IEA), in *Energy Technology Perspectives 2014* (2014).

[12] Thompson A., in *Major Greenhouse Gas Reductions Needed by 2050: IPCC*, April 13 (2014). Retrieved from: `http://www.climatecentral.org/news/major-greenhouse-gas-reductions-needed-to-curtail-climate-change-ipcc-17300`.

[13] Center for Climate and Energy Solutions (C2ES), *EPA Regulations of Greenhouse Gas Emissions from New Power Plants* (2015). Retrieved from: `https://www.c2es.org/federal/executive/epa/ghg-standards-for-new-power-plants`.

[14] World Resources Institute, in *CCS Overview: What Is CCS?* (2016). Retrieved from: `http://embarq.wri.org/project/carbon-dioxide-capture-storage/ccs-basics`.

[15] Global CCS Institute, in *The Global Status of CCS: 2015 Summary Report* (2015). Retrieved from: https://www.globalccsinstitute.com/publications/global-status-ccs-2015-summary-report.

[16] Global CCS Institute, in *Carbon capture and storage in the community of Portuguese language countries: opportunities and challenges* (2015). Retrieved from: https://hub.globalccsinstitute.com/publications/carbon-capture-and-storage-community-portuguese-language-countries-opportunities-and-challenges/32-sectors-potential-ccs-activities.

[17] Intergovernmental Panel on Climate Change, in *Carbon Dioxide Capture and Storage* (2005). Retrieved from https://www.ipcc.ch/pdf/special-reports/srccs/srccs_wholereport.pdf.

[18] International Energy Agency, in *The Way Forward: Five Key Actions to Achieve a Low-Carbon Energy Sector* (2014). Retrieved from: https://www.iea.org/publications/freepublications/publication/The_Way_forward.pdf.

[19] International Energy Agency, in *Prospects for CO_2 Capture and Storage* (2015). Retrieved from: https://www.iea.org/Textbase/npsum/ccsSUM.pdf.

[20] International Energy Agency, in *Technology Roadmap, Carbon capture and Storage* (2013). Retrieved from: https://www.iea.org/publications/freepublications/publication/technologyroadmapcarboncaptureandstorage.pdf.

EPJ Web of Conferences **148**, 00003 (2017) DOI: 10.1051/epjconf/201714800003
LNES 2016

Materials, critical materials and clean-energy technologies

R. EGGERT(*)

Division of Economics and Business, Colorado School of Mines - Golden, 80401 CO, USA

Summary. — Modern engineered materials, components and systems depend on raw materials whose properties provide essential functionality to these technologies. Some of these raw materials are subject to supply-chain risks, and such materials are known as critical materials. This paper reviews corporate, national and world perspectives on material criticality. It then narrows its focus to studies that assess "what is critical" to clean-energy technologies. The focus on supply-chain risks is not meant to be alarmist but rather to encourage attention to monitoring these risks and pursuing technological innovation to mitigate the risks.

1. – Introduction

Modern materials provide essential properties to engineered components and systems —properties such as luminescence, magnetism, thermal conductivity and corrosion resistance. Critical materials provide these and other essential properties and, moreover, are subject to supply-chain risks. Of particular concern in this paper are supply disruptions that might hinder the development and deployment of emerging clean-energy technologies such as electric vehicles, light-emitting-diode (LED) lighting, photovoltaic power systems and wind turbines.

Concern about raw-material availability, or rather limits on availability, is not new. It has been a recurring societal concern. In the middle 1800s, for example, W.S. Jevons

(*) E-mail: reggert@mines.edu

explored the exhaustion of coal in Britain [1]. In the late 1800s and early 1900s, dramatic increases in the use of fertilizers in commercial agriculture led to fears about the adequacy of naturally occurring sources of nitrogen, which at the time came from animal dung, urine, and Chilean nitrate deposits [2, 3]. In the years following World War II, concerns focused on a wide range of mineral raw materials and energy, given the significant use of mineral and energy resources during the war [4]. In the 1970s, attention returned to energy and minerals, in this case concerns about reliability of foreign sources of oil and some nonfuel minerals, as well as long-term adequacy of nonrenewable resources generally and implications for economic growth [5]. In each case, society adjusted to perceived scarcities through one or a combination of three approaches: developing new sources of mineral and energy resources, using resources more efficiently, and developing substitute materials or sources of energy.

The first decade or so of the 21st century witnessed a sustained boom in prices (since reversed) for many energy and mineral commodities, drawing public attention once again to resource availability. Part of this attention has focused on elements that, up until recently, have had limited commercial use. Typically, these elements represent a small part of the mass of an assembly or final product and yet provide essential functionality to the assembly or product —for example, neodymium and dysprosium in permanent magnets, europium and terbium in advanced lighting, indium and tellurium in photovoltaic materials, and lithium in batteries. The attention on these specialty metals led to several efforts to establish a framework for assessing what is "critical".

2. – Methodology

One of the first to suggest a methodology was a study by a committee of the U.S. National Research Council [6]. It presented a framework for assessing criticality in two dimensions: a) supply risk, or the probability of a supply disruption, and b) the impact of a supply disruption should it occur. Most subsequent studies represent variations on or extensions to this general framework.

The National Research Council (NRC) study notes that the criticality of specific elements changes over time as production technologies evolve and as new materials and products are developed. It also notes that, in the short to medium term, supply disruptions come largely from

- significant and unanticipated increases in demand, especially for elements with small existing markets (for example, rhenium for which annual world demand is on the order of several tens of tonnes),
- concentration of production in a small number of mines, companies, or countries, especially countries with significant political risks (such as cobalt and the geographic concentration of its production in central Africa), and
- production coming predominantly as a by-product (such as indium produced predominantly as a minor by-product of zinc metallurgical operations).

DOI: 10.1051/epjconf/201714800003

Over the longer term, mineral availability, or lack of availability, depends importantly on whether investments occur in mineral production and the underlying science and engineering that inform our understanding of mineral occurrence and define processes for efficient mineral exploration, mine development and mining and mineral processing.

The most comprehensive extension to the NRC methodology is Graedel *et al.* [7]. This paper adds a third dimension to criticality: environmental implications of material production and use. It also articulates important differences in perspective when considering corporations, nations and the world as a whole and defines indicators of criticality for all three dimensions and all three perspectives.

3. – The many faces of criticality

What is critical depends on the situation, as a number of recent studies illustrate — leading to a number of variations on the methodologies described above when assessing material supply chains in specific circumstances.

3·1. *Corporate perspectives.* – A corporate perspective asks, what materials are essential to a company's financial performance and what supply vulnerabilities put this performance at risk? Corporations are understandably reluctant to reveal their vulnerabilities when it comes to raw-material supply chains. Yet most presumably carry out some type of internal assessment of these vulnerabilities. General Electric (GE) is one of the few to have written publicly about their assessments [8, 9], which customize the general framework proposed in the NRC (2008) study discussed earlier. GE evaluated criticality in two dimensions: a) supply and price risk based on historical price volatility, sourcing and geopolitical risk, demand risk, ease or difficulty of substitution, co-production risk, and an element's abundance in the earth's crust; and b) impact on GE of a restriction based on ability to pass-through cost increases to customers, GE's ability to substitute, impacted GE revenue and GE's purchases as a percentage share of world production.

GE's strategy for ensuring raw-material availability consists essentially of the three approaches identified earlier: a) broadening material sources, b) optimizing use of materials through improvements in manufacturing efficiency and recycling and c) redesigning materials and systems, referred to as substitution earlier. Within the last decade, GE used a combination of these approaches to deal with two actual and one anticipated supply disruption. The first actual disruption involved rhenium used in superalloys in various types of turbines. In 2006, rhenium prices increased six-fold. In response, GE developed new relationships with suppliers, improved manufacturing efficiencies, developed new processes to reclaim and re-use rhenium in nickel-based superalloys from end-of-life turbine blades and developed and certified new alloys that contain less or no rhenium. The second actual disruption involved rare earths, which experienced ten-fold or greater price increases in 2010–2011. In response, GE is developing recycling technologies, especially for phosphor (lighting) materials, redesigning products to reduce rare-earth content (for example, in fluorescent lights) and developing new systems that use little or no rare earths (for example, LED lights that require fewer rare earths per unit of light). The anticipated supply disruption, which was averted, involved helium (^{3}He) used in neutron detectors, the demand for which was expected to increase significantly following the events of September 11, 2001. In response, GE worked with several entities to assure future supplies, developed a reclamation system for helium recycling and developed a boron-based system to replace helium-based systems for neutron detection.

3˙2. *National and regional perspectives.* – A national or (multinational) regional perspective is different than a corporate perspective. National and regional perspectives to date have focused on the vulnerabilities of national or regional manufacturing sectors to supply disruptions of important raw materials. Two studies illustrate this perspective.

European Commission [10] evaluates raw materials from the perspective of the European manufacturing sector. It evaluates criticality in two dimensions: a) supply risk based on substitutability, end-of-life recycling, and concentration of production in countries with poor governance; and b) importance of a raw material to European manufacturing, based on the gross value added of a sector. It evaluates fifty-four materials and finds that twenty exceed thresholds for supply risk and importance to qualify as critical.

Daisuke [11] assesses the perspective of the Japanese manufacturing sector using a modified version of the methodology used above for the European Union. The important modifications are incorporating the concentration by country of both reserves and Japanese import partners into the estimate of supply risk. Daisuke evaluates forty-one elements or groups of elements and, significantly, does not use this evaluation to create a list of critical materials; rather he simply displays relative degrees of supply risk and economic importance in a two-dimensional figure.

3˙3. *A world perspective.* – The broadest perspective is that of the world. Graedel *et al.* [12] present such a perspective, which additionally is noteworthy because of its methodology and timeframe. Of the studies considered here, Graedel *et al.* [12] is the only one that a) incorporates environmental implications for human health and ecosystems into its methodology and b) explicitly considers the long term, in this case 10–50 years into the future.

3˙4. *Comparisons.* – Tables I, II and III summarize the essential features of the major criticality assessments described above (scope and time frame, indicators of criticality and findings, respectively). They demonstrate the variety and diversity of analyses, suggesting to me, at least, that developing one definitive list of critical raw materials, applicable to all circumstances, is futile. Despite the common focus on criticality, these studies illustrate important differences in the following respects:

- *Elements or materials evaluated*: Which materials are evaluated necessarily reflects the perspective of the entity undertaking the evaluation.

- *Time frame (explicitly forward looking or not)*: Some studies represent a description of the present and thus are useful for evaluating the near future. Others are

TABLE I. – *Assessing criticality: scope and time frame. Sources: GE* [8,9], *Japan* [11], *European Union* [10], *World* [12].

	Scope of analysis	Time frame
Corporate		
General Electric (2011)	20 unspecified elements important to GE products, including 9 rare earths, emphasizing minor metals	Unspecified, implicitly a description of the present and thus relevant for up to about a decade into the future
National or Regional		
European Union (2014)	54 non-energy, non-agricultural materials (rare earths considered in three groups: light, heavy & scandium)	Unspecified, implicitly a description of the present and thus relevant for up to about a decade into the future
Japan (2015)	41 elements or groups of elements important for Japanese manufacturing (the element groups are heavy rare earths, light rare earths and platinum-group metals)	Unspecified, implicitly a description of the present and thus relevant for up to about a decade into the future
World		
Graedel *et al.* (2015)	62 metals and metalloids, including 15 rare earths	Long term, 10–50 years into the future

explicitly forward looking and evaluate what materials might become "at risk" in the medium or long term.

- *Methodology (indicators, quantitative or qualitative assessment)*: All studies capture to some degree the concepts of supply risk and the impact of a supply disruption should it occur. But the indicators used for supply risk and impact vary considerably from one study to another. Moreover, the degree to which these indicators are based on quantitative data or qualitative judgments also varies from case to case.

- *Presentation of findings*: Some studies label specific elements or materials as "critical", implying that others are not critical (an "either/or" view of criticality). Other studies, in contrast, embrace criticality as a "more or less" concept and present results that rank elements or materials according to their degree of supply risk and impact of a supply disruption should it occur.

DOI: 10.1051/epjconf/201714800003

TABLE II. – *Assessing criticality: indicators. Sources: GE* [8,9], *Japan* [11], *European Union* [10], *World* [12].

	Indicators of supply risk	Indicators of impact of a supply disruption
Corporate		
General Electric (2011)	Substitutability Demand risks Sourcing & geopolitical risks Historical price volatility Co-production risk Crustal abundance	Evaluated as impact on GE, based on: Impacted GE revenue, GE purchases as % of world production, Ability to pass through cost increases to customers, Substitutability
National or Regional		
European Union (2014)	Substitutability End-of-life recycling rates Concentration of primary production in countries with poor governance	Evaluated as economic importance to European Union manufacturers, based on: Gross value added of end-use megasectors
Japan (2015)	Substitutability End-of-life recycling rates Concentration of primary production in countries with poor governance Country concentration of reserves Concentration of import partner countries	Evaluated as economic importance to Japanese manufacturers, based on: Average import price, Quantity of domestic demand, Gross value added by consuming sectors
World		
Graedel *et al.* (2015)	Depletion time Co-production (% mined as a by-product)	Evaluated as vulnerability to a supply restriction, based on: Economic importance, Substitutability

– *Specific findings*: Not surprisingly given the differences noted above, each study comes to somewhat different specific findings. Nevertheless, a number of elements, or types of elements, appear as critical across the range of studies reviewed here: rare earth elements, and those produced largely as by-products of other, major materials (for example, cobalt, gallium, germanium, hafnium, indium, rhenium, and tellurium).

DOI: 10.1051/epjconf/201714800003

Table III. – *Assessing Criticality: findings. Sources: GE* [8,9], *Japan* [11], *European Union* [10], *World* [12].

	Reporting of findings	Specific findings
Corporate		
General Electric (2011)	No explicit "critical" list. Results displayed on a 2-dimensional figure.	Specific elements not identified, although 8 of the 9 rare earths appear among the relatively more critical elements
National or Regional		
European Union (2014)	20 materials identified as "critical" based on thresholds for supply risk and economic importance. Results also displayed on a 2-dimensional figure.	*Critical*: antimony, beryllium, borates, chromium, cobalt, coking coal, fluorspar, gallium, germanium, indium, magnesite, magnesium, natural graphite, niobium, platinum group metals, phosphate rock, heavy rare earths, light rare earths, silicon metal, tungsten
Japan (2015)	No explicit "critical" list. Results displayed on a 2-dimensional figure.	Among the relatively more critical: antimony, germanium, graphite, niobium, magnesium, platinum group metals, silicon, tungsten, heavy rare earths, light rare earths
World		
Graedel *et al.* (2015)	No explicit "critical" list. Results displayed in a 3-dimensional criticality space. Cluster analysis used to identify groups with similar degrees of criticality.	Specific results difficult to summarize. From the abstract: "The metals of most concern tend to be those available largely or entirely as byproducts, used in small quantities for highly specialized applications and possessing no effective substitutes."

4. – Technological perspectives and clean energy

Narrowing our focus, a technological perspective on criticality concentrates on the vulnerability of a particular technology to supply disruptions for important raw materials. For emerging technologies, it is the development and deployment of the technology that

TABLE IV. – *Potentially Critical Materials for Selected Clean-Energy Technologies. Sources:* [13-20] *and author's experience.*

Technology	Potentially critical materials
Batteries	
Lead acid	Lead
Nickel metal hydride	Lanthanum, nickel
Lithium ion	Cobalt, graphite, lithium
Emerging	Magnesium, vanadium
Solar power	
Silicon based	Nickel, silver, tin
Copper-indium-gallium selenide	Gallium, indium, selenium
Cadmium telluride	Tellurium
Magnets (for high efficiency motors)	
Alnico	Cobalt, nickel
Samarium-cobalt	Cobalt, samarium
Neodymium-iron-boron	Dysprosium, neodymium, praseodymium, terbium
Lighting	
Fluorescent	Cerium, europium, lanthanum, manganese, terbium, yttrium
Light-emitting diodes (LEDs)	Cerium, europium, gallium, germanium, indium, lanthanum, nickel, silver, terbium, tin, yttrium
Catalysts	Cerium, lanthanum, palladium, platinum, rhodium
Fuel cells	Cerium, cobalt, gadolinium, lanthanum, palladium, platinum, rhodium, yttrium
Nuclear	Cobalt, gadolinium, hafnium, indium
Hydrogen electrolysis	Palladium, platinum, rhodium

is at risk. For selected energy technologies, table IV summarizes potentially critical elements, many of which are the same as those identified by the broader criticality studies described above.

Two organizations have devoted considerable attention to identifying and prioritizing critical elements for clean-energy technologies, the U.S. Department of Energy and the European Commission's Joint Research Centre. Table V summarizes these studies.

4·1. *U.S. Department of Energy.* – Studies by the U.S. Department of Energy (DOE) in 2010 and 2011 use an extended version of the NRC framework to identify critical elements for selected materials in electric vehicles, fluorescent lighting, photovoltaic

DOI: 10.1051/epjconf/201714800003

TABLE V. – *Assessing criticality for energy technologies. Sources:* [14,16,17].

	Scope of analysis	Time frame	Indicators of supply risk	Indicators of impact of a supply disruption	Reporting of findings	Specific findings
US Department of Energy (2011)	16 elements used in wind turbines, electric vehicles, photo voltaic cells and fluorescent lighting, including 9 rare earths	Separate evaluations for the short term 2011–2015 and medium term 2015–2025	Projected market balance; competing energy demand; political, regulatory & social factors; co-production risk; producer diversity	Evaluated as importance to clean energy, based on: projected clean-energy demand & substitutability	5 elements identified as "critical" and 2 elements as "near critical" based on thresholds for supply risk and importance to clean energy. Results also displayed on a 2-dimensional figure.	*Critical*: dysprosium, europium, neodymium, terbium, yttrium *Near critical*: lithium, tellurium
European Joint Research Centre (2013)	Screening of 60 elements with important uses in wind, solar, carbon-capture-and-storage, nuclear -fission, bioenergy, electricity-grid, electricity-storage, electric vehicle, fuel-cell & lighting technologies. 33 examined in detail, including 10 rare earths.	Out to 2030	Evaluation based on 4 factors: (1) likelihood of rapid global demand growth, (2) limitations to expanding primary supply capacity, (3) concentration of primary supply, and (4) political risks of major supplying countries		Shortlisted materials placed into 5 categories: high, high-medium, medium, medium-low, low	*High*: dysprosium, europium, gallium, praseodymium, neodymium, tellurium, terbium, yttrium *High-medium*: germanium, graphite, hafnium, indium, platinum, rhenium
European Joint Research Centre (2016)	15 elements used in wind power, photovoltaics and electric vehicles	2015 and 2030	Evaluation based on 13 factors grouped into two categories (1) 8 upstream factors including material demand, investment potential, stability of supply, depletion of reserves, import reliance, supply adequacy, recycling and substitution and (2) 3 downstream factors including supply chain dependency, purchasing potential and material costs		EU resilience to material supply bottlenecks for each of 15 materials placed into High, Medium and Low categories for (1) 2015 and (2) for 4 scenarios in 2030 reflecting expected technology deployment and 4 scenarios reflecting varying degrees of mitigation	*For 2030, low resilience with no mitigation*: the rare earths dysprosium, nedodymium, praseodymium; cobalt; graphite; indium; lithium; silver *For 2030, low resilience even with mitigation*: lithium and the rare earths neodymium and praseodymium

power systems, and wind turbines [13, 14]. These studies evaluate supply risk on the basis of forward-looking calculations of the balance between production and demand for an element; competing technology demand; political, social and regulatory factors; co-production; and producer diversity (or lack thereof). These studies evaluate the second dimension of criticality, importance to clean energy, on the basis of likely growth in clean energy demand for an element, and ease or difficulty of substituting away from an element should a supply disruption occur. DOE evaluates sixteen elements and, in the 2011 study finds the rare earths dysprosium, europium, neodymium, terbium, and yttrium to be critical, with lithium and tellurium as near critical, looking forward five to fifteen years. In addition, these studies articulate a strategy for reducing supply-chain risks though innovation in three areas: process engineering to facilitate more efficient primary production, process engineering to increase manufacturing efficiency and recycling, and materials science and engineering to develop substitute materials.

4·2. *European Joint Research Centre.* – Studies published in 2011, 2013 and 2016 by the European Commission's Joint Research Centre (JRC) evaluate critical metals in the energy sector [15-17]. The 2011 and 2013 studies evaluate a broader set of technologies and materials than the U.S. DOE studies described in the previous section. The 2011 study encompasses the nuclear-fission, solar, wind, bioenergy, carbon-capture-and-storage, and electricity-grid sectors. The 2013 study includes the same six sectors and adds eleven technology sectors including fuel cells, electric vehicles, electricity storage and lighting. In the 2013 study, JRC evaluates sixty elements for possible detailed study and then studies thirty-two elements in some depth —using a methodology that considers likelihood of rapid demand growth, limitations on the ability to expand production capacity for an element experiencing demand growth, geographic concentration of production, and political risk. It identifies eight elements as critical: the rare-earth elements dysprosium, europium, neodymium, praseodymium, terbium, and yttrium, as well as gallium and tellurium. It identifies four materials as near critical: graphite, indium, platinum, and rhenium.

The 2016 JRC study is narrower in scope than the 2011 and 2013 studies, evaluating potential supply-chain bottlenecks for fifteen materials providing important functionality to systems in wind power, photovoltaics and electric vehicles. It assesses the situation in 2015 and then identifies potential bottlenecks in 2030 under forecasted deployment of these technologies under four scenarios that differ in the degree to which they incorporate measures to mitigate supply-chain risks —recycling, European Union (EU) production and substitution. With no mitigation, materials for which supply-chain resilience in the EU would be low in 2030 are: dysprosium, neodymium, praseodymium (wind power and electric vehicles), indium and silver (photovoltaics) and cobalt, graphite and lithium (electric vehicles only). With effective mitigation, supply-chain resilience in 2030 would be low only for neodymium, praseodymium and lithium. The JRC methodology in the 2016 study is noteworthy because it explicitly considers (separately) the upstream and downstream portions of the supply chains.

4·3. *Other energy studies.* – Several studies are narrower in scope or less detailed in their analysis of energy technologies and criticality.

A 2011 monograph by Achzet *et al.* [18], describes supply chains for nineteen elements or families of elements that provide important functionality in the oil, biomass, natural gas, coal, nuclear, geothermal, wind, hydro, and solar sectors. It subjectively assigns supply risk scores of high, medium, or low. The elements or families of elements receiving high-risk scores are: cadmium, chromium, cobalt, gallium, germanium, platinum, rare earths, rhodium, tellurium, and uranium.

A 2014 paper by Goe and Gaustad [19] evaluates critical materials for photovoltaics in the United States. It evaluates seventeen elements and, of these, finds six to be most critical: arsenic, germanium, indium, platinum, silver, and tin. The paper evaluates three types of risk: a) supply risk, based on net import reliance, concentration of production, recycling rates, and ratio of production to reserves; b) environmental risk, based on human toxicity, primary embodied energy, and energy savings; and c) economic risk, based on material price, domestic consumption, and economic value of the sector.

Another 2014 paper, this one by Roelich *et al.* [20] develops a methodology for evaluating supply risks dynamically —focusing on changes in criticality during the transition to electricity generation that produces lower carbon emissions than at present. The authors demonstrate their methodology by applying it to neodymium and the power sector of the United Kingdom.

5. – What to do?

As noted earlier, there are three obvious responses that have the potential to mitigate criticality: 1) increase and diversify primary production, 2) enhance the efficiency with which we use materials (increased manufacturing efficiency, material re-use and recycling) and 3) develop substitute materials. From the perspectives of governments and public policy, however, the central issue is determining who should be responsible for undertaking these responses.

Natural market forces provide powerful incentives for material users and producers, most of whom are not governments, to respond to criticality. Thus my personal view is that the primary responsibility rests with material users and producers, including potential producers. Users, typically manufacturers of components and assemblies destined for final products, should act in their own self interests to evaluate and secure their supply chains —in effect, buying insurance against supply-chain disruptions through one or a combination of activities such as entering strategic alliances with material suppliers, diversifying sources of supply, and over the longer term perhaps developing substitute materials. Similarly, as long as markets properly signal supply-chain risks, then material producers (actual and potential) will act to develop new and more diversified sources of supply by investing in mineral exploration, mine development, and technological research that improve material supplies.

Public policy, in turn, should focus on activities that enhance the functioning of markets, especially when markets do not work well on their own. Specifically, as noted in Eggert [21, 22]:

- Facilitating international trade when trade restrictions are a significant source of supply-chain risk.
- Facilitating development of new mines and processing facilities when undeveloped resources are known to (or may conceivably) exist.
- Investing in the provision of basic market information and strategic analysis, on which actual material users and producers can make informed decisions about critical materials (a key difficulty that market participants face in making informed decisions about many critical materials is, arguably, that many of critical-material markets are small and opaque —that is, difficult to evaluate on the basis of public information); and
- Investing in research and education over the entire material life cycle.

Information, research, and education are, at least in part, what economists call public goods, in that the private sector alone is likely to underinvest in these activities from society's perspective because the benefits of these investments are risky, far in the future for the most part, and difficult for private investors to fully appropriate.

6. – Conclusions

Criticality is in the eye of the beholder and has many faces. The perspective of a corporation necessarily differs from that of a nation or region's manufacturing sector, which, in turn, differs from those of an emerging technology or the world as whole. Moreover, criticality is dynamic; what is critical today likely will not be critical in the future because of actions taken today to thwart vulnerability to supply restrictions for important raw materials. Criticality assessment is an important tool for identifying potential problems before they occur.

* * *

This paper draws significantly on two earlier papers published in conference proceedings [23, 24]. This work overall is supported in part by the Critical Materials Institute, an Energy Innovation Hub funded by the U.S. Department of Energy, Office of Energy Efficiency and Renewable Energy, Advanced Manufacturing Office, and the Viola Vestal Coulter Foundation.

REFERENCES

[1] Jevons W. S., *The Coal Question: An Inquiry Concerning the Progress of the Nation, and the Probable Exhaustion of Our Coal Mines* (Macmillan & Co, London and Cambridge) 1865.
[2] Hager T., *The Alchemy of Air: A Jewish Genius, A Doomed Tycoon, and the Scientific Discovery That Fed the World But Fueled the Rise of Hitler* (Random House, New York) 2008.

[3] Smil V., *Enriching the Earth: Fritz Haber, Carl Bosch, and the Transformation of World Food Production* (MIT Press, Cambridge, Massachusetts) 2004.
[4] President's Materials Policy Commission, *Resources for Freedom*, Vol. **1**: *Foundations for Growth and Scarcity* (U.S. Government Printing Office, Washington, DC) 1952.
[5] Smith V. K. (Editor), *Scarcity and Growth Reconsidered* (Johns Hopkins University Press for Resources for the Future, Baltimore, Maryland) 1979.
[6] National Research Council, *Minerals, Critical Minerals, and the U.S. Economy* (National Academies Press, Washington, DC) 2008.
[7] Graedel T. E., Barr R., Chandler C., Chase T., Choi J., Christoffersen L., Friedlander E., Henly C., Jun C., Nassar N. T., Schechner D., Warren S., Yang M. and Zhu C., *Environ. Sci. Technol.*, **46** (2012) 1063, doi:10.1021/es20354z.
[8] Duclos S., *Sustainable materials for manufacturing: GE risks and opportunities*, paper presented at the *GE Whitney Symposium, Niskayuna, New York, October 2011*.
[9] Ku A. and Hung S., *CEP Magazine*, September 2014, pp. 28–35.
[10] European Commission, *Report on critical raw materials for the EU*, retrieved from `http://ec.europa.eu/growth/sectors/raw-materials/specific-interest/critical/index_en.htm`, 2014.
[11] Daisuke A., *A study of a stable supply of mineral resources*, unpublished study of the Japan Oil, Gas and Metals National Corporation, Tokyo, 2015.
[12] Graedel T. E., Harper E. M., Nassar N. T., Nuss P. and Reck B. K., *Proc. Natl. Acad. Sci. U.S.A.*, **112** (2015) 4257, doi: 10.1073/pnas.1500415112.
[13] Department of Energy, *Critical Materials Strategy* (U.S. Department of Energy, Washington, DC) December 2010.
[14] Department of Energy, *Critical Materials Strategy* (U.S. Department of Energy, Washington, DC) December 2011.
[15] Moss R. L., Tzimas E., Kara H., Willis P. and Kooroshy J., *Critical Metals in Strategic Energy Technologies* (European Union, Luxembourg) 2011.
[16] Moss R., Tzimas E., Willis P., Arendorf J., Tercero Espinoza L. *et al.*, *Critical Metals in the Path towards Decarbonisation of the EU Energy Sector: Assessing Rare Metals as Supply-Chain Bottlenecks in Low-Carbon Energy Technologies*, JRC Scientific and Policy Report (European Union, Luxembourg) 2013.
[17] Blagoeva D. T., Aves Dias P., Marmier A. and Pavel C. C., *Assessment of potential bottlenecks along the materials supply chain for the future deployment of low-carbon energy and transport technologies in the EU*, JRC Science for Policy Report (European Union, Luxembourg) 2016.
[18] Achzet B., Reller A. and Zepf V., *Materials Critical to the Energy Industry: An Introduction* (University of Augsburg, Augsburg) 2011.
[19] Goe M. and Gaustad G., *Appl. Energy*, **123** (2014) 387.
[20] Roelich K., Dawson D. A., Purnell P., Knoeri C., Revell R., Busch J. and Steinberger J. K., *Appl. Energy*, **123** (2014) 378.
[21] Eggert R. G., *Issues Sci. Technol.*, **XXVI** (2010) 49.
[22] Eggert R. G., *Nat. Chem.*, **3** (2011) 688.
[23] Eggert R. G., *Proceedings of the World Engineering Conference and Convention, Kyoto, Japan, 2015*, paper 20533, oral session 3-2-1.
[24] Eggert R. G., *XXVIII International Mineral Processing Congress Proceedings, Quebec City, Canada, 2016* (The Canadian Institute of Mining, Metallurgy and Petroleum) 2016, paper 739.

Energy in buildings: Efficiency, renewables and storage

Matthias M. Koebel(*)

Swiss Federal Laboratories for Materials Science and Technology, Empa
Ueberlandstrasse 129, CH-8600, Duebendorf, Switzerland

Summary. — This lecture summary provides a short but comprehensive overview on the "energy and buildings" topic. Buildings account for roughly 40% of the global energy demands. Thus, an increased adoption of existing and upcoming materials and solutions for the building sector represents an enormous potential to reduce building related energy demands and greenhouse gas emissions. The central question is how the building envelope (insulation, fenestration, construction style, solar control) affects building energy demands. Compared to conventional insulation materials, superinsulation materials such as vacuum insulation panels and silica aerogel achieve the same thermal performance with significantly thinner insulation layers. With low-emissivity coatings and appropriate filler gasses, double and triple glazing reduce thermal losses by up to an order of magnitude compared to old single pane windows, while vacuum insulation and aerogel filled glazing could reduce these even further. Electrochromic and other switchable glazing solutions maximize solar gains during wintertime and minimize illumination demands whilst avoiding overheating in summer. Upon integration of renewable energy systems into the building energy supply, buildings can become both producers and consumers of energy. Combined with dynamic user behavior, temporal variations in the production of renewable energy require appropriate storage solutions, both thermal and electrical, and the integration of buildings into smart grids and energy district networks. The combination of these measures allows a reduction of the existing building stock by roughly a factor of three —a promising, but cost intensive way, to prepare our buildings for the energy turnaround.

(*) Email: `matthias.koebel@empa.ch`, `www.empa.ch`

Fig. 1. – Modern energy-efficient buildings require a balanced interplay between an efficient envelope (insulation), the on-site generation of renewables and energy storage in a suitable form.

1. – Overview and context

Our built environment is the largest energy consuming sector —roughly 40% of the world's energy demand is attributed to heating, ventilation and air conditioning (HVAC) of buildings [1]. Modern buildings perform substantially better because of tightening building codes and regulations regarding their energy performance, nevertheless the majority of existing buildings are older and thus in dire need for refurbishment. The three key steps to a low-energy or plus-energy building are i) an efficient envelope which minimizes thermal losses and maximizes solar gains in winter while excluding them in the summer, ii) harvesting of renewable energies at the building itself (*e.g.* through photovoltaics) sufficient to cover a large fraction of the buildings annual consumption and iii) putting in place suitable energy storage systems (thermal or electrical). Furthermore, thermal and electrical storage systems can be shared between multiple houses within a district in order to manage and balance the dissimilar time-dependent production/demand profiles of individual homes (*e.g.* through smart IOT solutions). Such energy district network concepts allow significant energy savings at district and city level operation and reduce peak loads on the energy infrastructure (fig. 1).

2. – Thermal insulation and superinsulation materials

To address the first problem, namely improving the quality of the building envelope to reduce thermal losses, let us consider the installation of suitable insulation materials. Insulation performance of a material is described in terms of the thermal conductivity λ with units of W/(m K). Typical conventional materials include foamed plastics

(Polystyrene foam, $\lambda \sim 0.03\,\mathrm{W/(m\ K)}$, PU/PUR/PIR foam, $\lambda \sim 0.024\,\mathrm{W/(m\ K)}$), mineral and glass wool ($\lambda \sim 0.033\,\mathrm{W/(m\ K)}$) and a number of other less commonly used materials (cellulose fibers, natural materials, melamine foam etc.), see table I. All those materials are highly optimized and offer good value for money but are close to their respective product life cycle and can no longer keep up with ever increasing demands of European building codes.

Architects and planners are generally not thrilled with current building policy which essentially calls for ever increasing thicknesses of insulation layers. But are there high-performance alternatives to existing insulation materials which allow slimmer construction? Sure there are. To understand what they are and how they work, we need to first understand the limitations of existing insulation materials. In a thermal insulator, air is responsible for the majority of total heat transport. In a porous insulation material with submillimeter pore sizes, the total thermal conductivity λ_{tot} is the sum of gas, solid and radiative conductivities, respectively according to

$$\lambda_{\mathrm{tot}}(T,p) = \lambda_{\mathrm{g}}(T,p) + \lambda_{\mathrm{s}}(T) + \lambda_{\mathrm{r}}(T). \tag{1}$$

Both solid and radiative conductivity are more or less independent of pressure. Note that for the discussion of building insulation, people generally refer to ambient temperature and pressure conditions, although many industry standard measurements are done at 10 °C. For still ambient air, the gas conductivity, $\lambda_{\mathrm{g},\infty}$ is roughly $0.026\,\mathrm{W/(m\ K)}$ and is clearly the dominant form of heat transport in normal insulation materials by comparison with the above insulation material λ values. This shows that skeletal and radiative transport in these materials typically adds up to 3–10 mW/(m K) depending on the structure. It also illustrates the intrinsic difficulty for further reducing gas transport with ambient air as a filler gas. Polyisocyanate foams such as PU/PUR/PIR owe their high performance to a blowing gas mixture with lower gas conductivity (typically mixtures of CO_2, $\lambda_{\mathrm{g}} = 0.015\,\mathrm{W/(m\ K)}$ and isopentane, $\lambda_{\mathrm{g}} = 0.0144\,\mathrm{W/(m\ K)}$). To further decrease lambda values, halogenated (*e.g.* fluoro-chloro)-hydrocarbons have been used in the past, but these have been banned due to their ozone layer harming potential. Yet there are alternatives to achieve a further reduction of the gas conductivity. A further reduction of the gas conductivity into the so-called superinsulating regime (recent definition: $\lambda < 0.02\,\mathrm{W/(m\ K)}$ [2]) without the use of harmful or trace noble gas fillers (Kr, Xe) can also be achieved by other means:

$$\lambda_{\mathrm{g}} = \lambda_{\mathrm{g},\infty}\frac{1}{(1+2\beta Kn)}, \tag{2}$$

$$Kn = \frac{l}{\Phi}. \tag{3}$$

The gas conductivity for a gas where the predominant mode of energy transport is collisions between gas molecules is referred to as the so-called free gas conductivity $\lambda_{\mathrm{g},\infty}$. Equation (2) shows a correction term to the free gas conductivity which is inversely proportional to the Knudsen number Kn [3]. β is a gas-dependent constant which is about

TABLE I. – *Overview of standard building envelope technologies and solutions, particularly for thermal insulation and for windows.*

	Thermal conductivity [W/(m K)]	Combustible	Material cost [EUR/m^3]	U-value [W/(m^2 K)]	g-value []	Glazing cost EUR/m^2	Service life [a]
Conventional insulation							
Mineral wool	0.032–0.045	no	120				> 50a
Glass wool	0.030–0.040	no	120				> 50a
Polystyrene foam	0.029–0.037	yes	70				> 50a
PUR/PIR foam	0.021–0.025	yes	160				> 30a
Superinsulation							
Silica aerogel blanket	0.015–0.018	no	3000				> 50a
Silica aerogel granulate	0.018–0.022	no	2800				> 50a
Silica aerogel powder	0.020–0.025	no	2500				> 50a
Fumed silica core VIP	0.007	no	4500				> 30a
Window technology							
Standard double glazing (soft low-e, Ar fill)				1.0–1.1	0.6–0.8	45–60	> 25a
Standard triple glazing (soft low-e, Ar fill)				0.65–0.75	0.45–0.6	65–80	> 25a
Vacuum glazing (soft low-e, high vacuuml)				0.3–1.1	0.6–0.8	120–250	??

1.6 for air. The Knudsen number is defined as the ratio of mean free path of the filler gas l to the critical container length Φ (eq. (2)). For insulation materials, the latter can be approximated by the pore size. At ambient pressure, the mean free path of air molecules is in the 70–100 nm range. For typical insulation materials with pore sizes in the hundreds of microns, the Knudsen number is then on the order of 10^{-3} which makes the correction term to the free gas conductivity become very close to 1. However if we were remove the air from an insulator by partial evacuation, the correction term would strongly differ from unity. At 1mbar vacuum, the mean free path of air is also on the order of 100 microns which would yield a Knudsen number of 1. In this case, one would calculate a gas conductivity of roughly $0.026\,\mathrm{W/(m\,K)} \times 1/4.2$ which is approximately $6\,\mathrm{mW/(m\ K)}$ or roughly $20\,\mathrm{mW/(m\ K)}$ lower than that of free air —a massive reduction. The corresponding product is a vacuum insulation panel (VIP), commercially available in the form of a porous core (fumed silica is the material of choice for building applications) and a high-tech metallized multilayer laminate foil to ensure gas tightness [4]. Such VIPs attain declaration values of 0.006–$0.007\,\mathrm{W/(m\ K)}$ at vacuum levels in the single digit mbar range and service life numbers up to 30 years. However, they are subject to aging, particularly at temperatures $> 50\,^{\circ}\mathrm{C}$ and high humidity and of course to mechanical damage (piercing of the foil) which renders them less attractive for building insulation applications.

A second way to significantly lower the gas conductivity is to reduce the pore size of the material to the nanoscale size range. Such superinsulating materials are aerogels with typical average pore sizes on the order of 40 nm. Using again eq. (2) and the Knudsen number for ambient air ($40\,\mathrm{nm}/70\,\mathrm{nm} \sim 0.57$), the gas conductivity is reduced from 26 to $9.2\,\mathrm{mW/(m\ K)}$. Aerogels are a class of highly porous materials consisting of a fine pore structure with a majority of pores in the mesopore range (2–50 nm). They exist in many different compositions, chemistries (inorganic, organic, metallic) and building block morphologies (single phase, hybrids, hierarchical) [5]. The most common type of aerogels are based on silica and the only commercial materials today (see fig. 2). Monolithic products are difficult and costly to manufacture and have not yet been commercialized although BASF is currently working on a polyurea-based aerogel monolith product which could become commercially available as early as 2019 [6]. Blanket and granulate products are available from leading manufacturers Aspen Aerogels and Cabot. For building applications, these semifinished products must be adapted to solutions and products which are commensurate with the construction process. Aspen blankets are glued together and cut to create insulation boards such as Heck Aero which then finds application in aerogel-based ETICS solutions. Wet applied insulating render [7] and concrete [8] products are based on aerogel granulate.

Since the turn of the millennium, many demonstration objects and solutions have been realized using superinsulating materials and components. However these are primarily niche applications such as flat roof balcony insulation, window reveals, roller blind shutter box insulation and alcoves. Despite the high performance and added value of these solutions, price is still the key inhibitor for rapid market uptake, particularly for aerogel materials. If we want to achieve broader market uptake of aerogel-based solutions, novel production concepts need to be found and implemented industrially.

Current pricing of semi-finished aerogel products (granulate and blankets) is between 2500 and $3000\,\mathrm{EUR/m^3}$, however, the cost for the installed solution for the end user is roughly twice that. At current pricing level, the additional benefit of an aerogel solution is annulled by the high pricing leading to poor market acceptance and effectively condemning aerogels to niche markets in the B&C industry.

 DOI: 10.1051/epjconf/201714800004

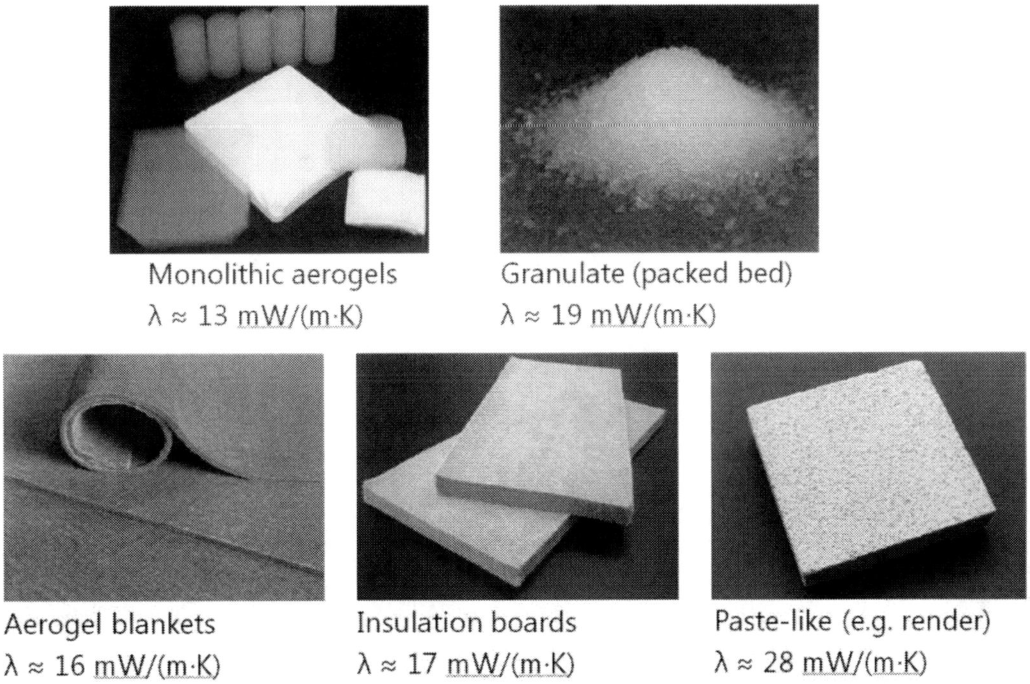

Fig. 2. – Aerogel superinsulation materials offer promising technological solutions for building retrofit and new construction in urban aereas.

Experts believe that prices must be roughly cut in half to reach a more significant market penetration of superinsulating solutions in the built environment. For aerogels this means raw materials and installed product cost of roughly EUR $1500/\mathrm{m}^3$ and 2500–3000 $\mathrm{EUR/m}^3$, respectively. In order to reach these ambitions targets, new industrial processes must be developed that allow faster and more economical production of aerogels allowing for further price reductions through economy of scale. In parallel, insulation system providers and aerogel expert groups need to work together to develop new solutions which are easier to install and more user-friendly. Last but not least, innovative products and solutions must be made visible to the public in the form of lighthouse projects ideally supported by prominent media coverage in order to increase awareness. In this way, superinsulating materials can develop into a valuable industry for applications in retrofitting and urban construction.

3. – Windows and glazing technology

The second central pillar of building envelopes are the transparent part, namely glazing. In Europe, current glazing standards feature noble gas filled double and triple glazing with low-emissivity (low-E) coatings reaching typical center of glazing U-values of 1 and $0.65\,\mathrm{W/(m^2\ K)}$, respectively (see table I). This may seem rather high in comparison with new envelope insulation standard which are on the order of 0.2–0.3 W/(m^2 K), however windows generate significant solar gains also during winter months which normally outweigh thermal losses. Next generation technologies in the glazing industry are vacuum and multichamber glazing. A vacuum insulation glazing (VIG) is a double glazing unit with a high vacuum in between which completely eliminates gas conduction [9]. To prevent the two panes from touching under the load of the atmospheric pressure (10 tons/m^2!), an array of thin spacers —so-called pillars— maintains the vacuum gap which is typically 0.2–0.7 mm thin. Theoretically, good vacuum glazing can achieve center of glazing U values as low as 0.3 W/(m^2 K) whit a total glazing thickness of less than 1 cm. However the edge seal of VIGs is still a challenge to commercialize industrially and existing products today feature higher U-values which partly mitigates the problem of mechanical strains under thermal loads which is critical for the system's service life. Modern low-U-value alternatives to VIG currently under development are multichamber glazings ($>$ 4 chambers) with either thin glass ($<$ 1.5 mm) or flexible, transparent polymer foils as separators. Such systems can achieve U-values as low as 0.3 W/(m^2 K) with weight comparable to a conventional triple glazing.

Aside from reducing thermal losses, modern window/glazing concepts also target the development of solar control technologies. Overheating during the hot summer months often represents a serious issue which typically leads to significantly increased cooling loads and thus electrical energy demand. In a standard double/triple glazing, the total solar energy transmittance or g-value is between 0.5 and 0.75 meaning that these windows transmit between 50 and 75% of incident total solar energy. Static (constant g-value) solar control measures include solar protection glass which contains selective coatings that only transmit the visible part of the optical spectrum but reflect infrared radiation and tinted glass. Dynamic solutions respond to an external stimulus and are promising alternatives to mechanical blinds which are maintenance intensive and sensitive to storm/hail damage. Modern concepts of such dynamic solutions include micro-vane blinds such as the MicroShade product by the Danish company Photosolar which features angular dependent g-values (low solar gains in summer, high solar gains in winter). Other solutions such as thermochromic or electrochromic windows alter their transmittance from dark to transparent in response to a temperature change or an external electrical control signal. Although very promising in the long-term, such smart window concepts are still rather expensive and need additional technical improvements (faster switching times, higher contrast, improved durability) in order to become attractive for large volume application.

4. – Renewable energy integration —our home as a powerplant

With an efficient building envelope, the energy demand of a building decreases by roughly 85% from 200 kWh/(m^2 a) for a non-insulated building to approximately 30 kWh/(m^2 a). With such a low energy demand, integration of renewable energy, mostly PV and solarthermal, is an economic way to cover the entire demand of a building.

Photovoltaic cells (PV) have made tremendous technological progress over the last three decades and prices have dropped accordingly, following roughly Moore's law which

Fig. 3. – Left: oldest solar grid-connected PV system in Europe at the University of applied Sciences near Lugano, CH. Right: retrofitted plus energy building in Romanshorn, CH featuring a high-performance BiPV façade, courtesy Viriden architects.

is well known in the related silicon-based semiconductor industry. Various types of cells exist, however by far the most common type are silicon-based single junction cells with commercial products reaching 20% conversion efficiency at module prices below 0.50 EUR per W peak power [10]. Solar thermal energy harvesting by means of thermal collectors is a parallel technology which is relatively mature in Europe but has the drawback a large amount of heat goes unused during warm summer months when demand bottoms out. Two prominent examples of a historical PV system and a building integrated photovoltaics (BIPV) retrofit are shown in fig. 3. The left example is the oldest grid connected PV system installed at the University of Applied Sciences SUPSI at Manno near Lugano.

After 30 years of continuous operation of this installation, those early silicon PV modules with a design power of 10 kWp still have a performance > 85%. The modern looking building on the right is a retrofitted 1960s multifamily home in Romanshorn, Switzerland which was converted to plus energy by means of a state of the art solar façade. Both examples clearly demonstrate that buildings represent a tremendous potential for the future integration of renewables into our energy infrastructure. Other forms of renewable energies such as wind, hydropower and biofuels typically require larger installations to operate efficiently and thus are not well-suited for building integration. With rapidly decreasing PV pricing and pending the large volume availability of electrical storage solutions such as stationary batteries or EV parks accessible through intelligent internet of things (IOT) solutions, it is expected that buildings are going to become a major producer of renewable energy in the coming decades. The main advantage of BiPV is that some of the installation cost can be synergistically compensated during the construction period of the façade and roof, whereas the rest is amortized during operation within typically 5–7 years.

5. – Energy storage and thermal energy upgrading

The last step towards a sustainable plus energy building (see fig. 1) is the integration of suitable storage and conversion solutions. Electrical energy storage in a home should be designed to buffer primarily short to medium term (hours to days) energy fluctuations expected from PV systems, thus maximizing in-house utilization of the renewable production. A number of commercial battery storage solutions exist, the oldest and cheapest technology being lead acid battery technology. However this type of storage offers relatively poor long-term performance (so-called cycle stability) and as a result is not ideally suited for stationary applications. More promising concepts include Nickel metal hydride (NiMH) and Li-ion battery technology, with the latter offering the advantage of higher power density and attractive cost perspectives (as low as 200 EUR/kWh capacity) due to massive manufacturing capacity extension as a result of a rapid acceleration of electric vehicle (EV) technologies worldwide. The disruptive EV manufacturer Tesla Motors has recently commercialized a battery storage system for home installation with an upgraded 14 kWh capacity per module.

Other battery concepts for larger capacity station (primarily suited for district networks and as buffers for solar farms) are based on high-temperature battery systems (sodium sulfur battery, zebra battery, thermal battery). A related, promising concept is the so-called liquid-metal battery developed by MIT researchers which promises extremely high cycle stability and low system cost. Commercialization of these modules is pending until the successful long-term testing.

Electrical long-term storage is not really a practical solution to cover the production shortage and increased demand of buildings, particularly in colder areas with elevated heating demand. Thermal energy on the other hand can be stored seasonally in a much simpler manner. This means that a substantial fraction or ideally the complete heating demand during the cold season is generated in the summer months, for example from solarthermal collectors, and then gradually released into a low temperature "low exergy" floor heating system. This would allow avoiding the consumption of precious high-value energy forms such as fossil fuels or electricity to cover the heating demand. The classical approach for seasonal heat storage is to use sensible heat *i.e.* heat up large volumes of a medium with a high heat capacity such as water. However such a technical solution demands tremendous water storage reservoirs. Figure 4 shows a 205 m^3 hot water storage reservoir needed to supply the winter heating demand for an 8-family zero-energy apartment building in Burgdorf, Switzerland realized by Jenni Energietechnik AG during the construction period. Alternative to sensible heat, latent heat storage system employ the typically much higher thermal energy density during phase transitions (melting/solidification) or chemical reactions (hydration reactions of crystalline materials) and thus require much less space for the same job. Such a system which is under development at Empa is a high ettringite content seasonal storage concrete which is dried using solar energy in the summer months and releases its stored energy upon rehydration (water addition). For a typical single-family home with a high quality envelope, it would require about 7–10 m^3 of this concrete to cover the winter heating demand. A first prototype building has recently been equipped with this technology.

An even more promising class of latent-heat–based thermal systems is the thermally driven sorption heat pump technology. Such devices are able to transform low-grade ($< 100\,°C$) waste heat into more precious forms of thermal energy. Early concepts in the

Fig. 4. – Seasonal thermal storage solutions based on sensible and latent heat, respectively.

1960s and 1970s were based on liquid sorbents with the disadvantage of high sensible heat losses (*e.g.* alcohol-water, ammonia). Modern systems are using the adsorption principle, which means they employ solid sorbents with a significantly lower cycled thermal mass. In comparison with a conventional heat pump in which a mechanical compressor drives the cycle, adsorption heat pumps make use of adsorption/desorption cycles of a sorbate, most often water, onto a porous solid sorbent to generate a temperature lift (or drop). One of the most promising applications from a global perspective is the transformation of waste or solarthermal energy into cooling energy, the latter often being referred to as solar cooling. Current commercial machines show already decent performance and are reliable; however improvements at both, the sorbent materials and the device and heat exchanger level can still bring significant improvements and increase competitiveness of the technology. Innovative manufacturers such as Sortech in Germany have recently developed the so-called hybrid compressor-driven/sorption heat pump machines which offer improved flexibility and exceptional performance under partial cooling loads. Figure 5 shows a schematic representation of a thermally driven heat pump at system level. Typical thermal heating and cooling coefficients of performance (COPs) are on the order of 1.5 and 0.6, respectively. Since the overall heat pump system still requires minimal amounts of electrical energy (control electronics, valves, pumps) a second figure of merit is the respective electrical COP. These are in the order of 40 and 15 for heating and cooling, respectively. For Switzerland, the heating and cooling demand of 112 and 28 TWh, respectively could be covered to a large extent from waste and solarthermal heat sources if sorption heat pump technology were to replace compressor-driven systems [11]. Even though thermal storage and adsorption cooling are poorly known and implemented, technologies focusing on more efficient usage of waste heat such as seasonal storage and thermally driven heat pumps are expected to play an important role in energy networks. Interconnecting multiple buildings with dissimilar energy demand, production and user behavior through energy hub and energy district network systems at neighborhood, district and urban scale will be able to create excellent synergies, effectively allowing energy

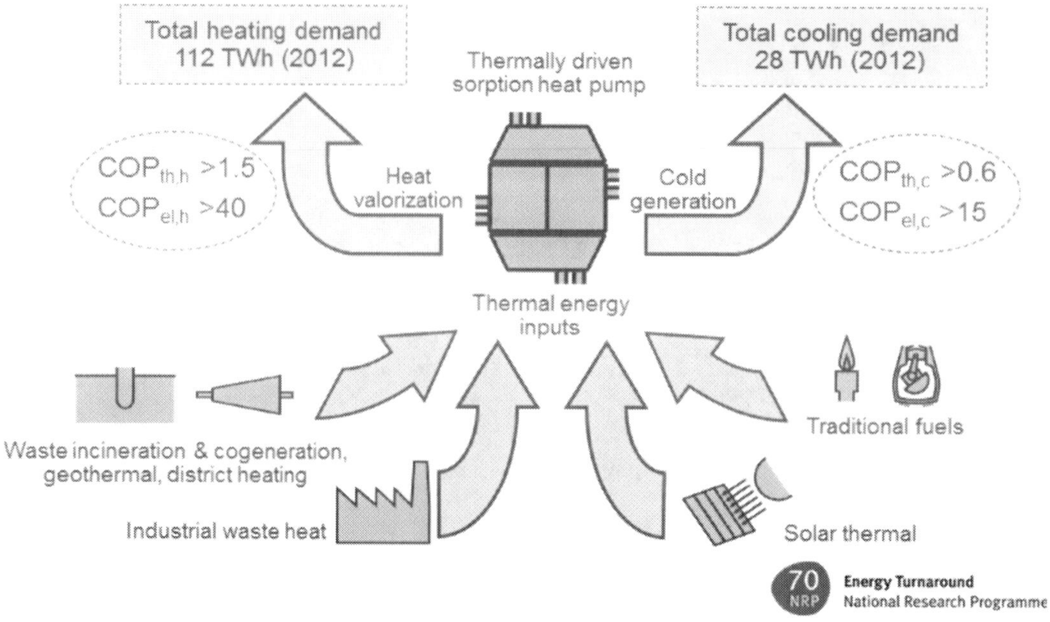

Fig. 5. – Principle of operation and key performance figures of thermally driven solid sorption heatpumps which offer a tremendous electrical energy savings potential through the valorization of low-grade waste heat.

to be used where it is created and thus reducing conversion, transport and storage losses. This is the last and important link that connects individual buildings to create the future, energy efficient cities.

6. – Need for technology demonstration and outlook

Clearly one of the main shortcomings of the construction sector is its conservative nature and poor innovation dynamics. As a result, most home owners, investors, architects, planners and craftsmen opt for well-established conventional products. Together with certification requirements, this imposes a substantial barrier for new innovative products such as aerogel superinsulation, smart windows or thermally driven sorption heat pumps. Hence one needs to create demonstrator platforms to increase visibility of such solutions. The goal is to increase awareness and educate people about new possibilities. In this way, one can reduce time to market for sustainable technologies, systems and processes in the area of building, mobility, energy and water. NEST is a new living lab [12] at Empa in Dübendorf, Switzerland which aims to combine all state-of-the-art solutions in the building, construction and energy sector. On the monitoring side, it employs both energy fingerprinting of individual units (small apartments or office segments and direct user feedback on comfort, thus providing essential feedback for the industry during product testing and market introduction.

In summary, the building energy topic consists of many different technological elements, none of which by themselves can solve the overall problem *i.e.* create low- and zero-energy buildings. As illustrated in the introductory fig. 1, economical solutions combine efficient envelopes with renewables and storage solutions. Applied research needs to fuse with innovative industrial R&D, manufacturing and marketing teams to create the solutions of the future which we all depend on.

REFERENCES

[1] Perez-Lombard L., Ortiz J. and Pout C., *Energy Build.*, **40** (2008) 394.
[2] Koebel M. M., Rigacci A. and Achard P., *J. Sol-Gel Sci. Technol.*, **63** (2012) 315.
[3] Fricke J. and Borst W. L., *Essentials of Energy Technology: Sources, Transport, Storage, Conservation* (Wiley VCH) 2013, Chapt. 2, ISBN: 978-3-527-33416-2.
[4] Baetens R., Jelle B. P., Thue J. V. *et al.*, *Energy Build.*, **42** (2010) 147.
[5] Aegerter M. A., Leventis M. and Koebel M. M., *Aerogels Handbook* (Springer Publishing) 2011, ISBN: 978-1-4419-7477-8.
[6] `https://www.basf.com/en/company/news-and-media/news-releases/2016/07/p-16-248.html`.
[7] `http://www.fixit.ch/aerogel/?w=start&lng=en`.
[8] Ratke L., *Beton Stahlbetonbau*, **103** (2008) 236.
[9] Collins R. E., Turner G. M., Fischer-Cripps A. C., Tang J.-Z., Simko T. M., Dey C. J., Clugston D. A., Zhang Q.-C. and Garrison J. D., *Build. Environ.*, **30** (1995) 459.
[10] Ito M., Lespinats S., Merten J., Malbranche P. and Kurokawa K., *Prog. Photovoltaics*, **24** (2016) 159.
[11] `http://www.nfp70.ch/en/projects/building-settlement/heat-utilisation-with-sorption-technology`.
[12] `http://www.empa.ch/web/nest/`.

EPJ Web of Conferences **148**, 00005 (2017) DOI: 10.1051/epjconf/201714800005
LNES 2016

Power system and market integration of renewable electricity

Georg Erdmann(*)

Energy Systems of TU Berlin - Einsteinufer 25, D-10587 Berlin, Germany

Summary. — This paper addresses problems of power generation markets that arise under high shares of intermittent generation. After discussing the economic fundamentals of wind and photovoltaic investments, the paper introduces the concept of the "Merit order effect of renewables". According to this concept electricity prices on wholesale power markets become smaller in periods during which large volumes of wind and photovoltaic generation is available and squeeze out relative expensive gas-fired power generation. The merit order effect of renewables has a couple of consequences. Among others it challenges the profitability of conventional power generation. If such generation capacities are still necessary, at least during a transitory period, a capacity mechanism may be put in place that generates an additional stream of income to the operators of conventional power generators. Another consequence of growing intermittent power generation is the need for concepts and technologies that deal with excess generation. Among these concepts are virtual and physical power storage capacities. In the last parts of the paper models are presented that are able to analyze these concepts from an economic point of view.

1. – Introduction

The last years have seen a strong increase in renewable-electricity generation in many countries. In particular capacities of onshore wind, offshore wind and solar generation are everywhere on the rise. The global potentials of these renewable technologies are large and sufficient to cover the total electricity demand if these technologies could be dispatched according to the schedule of electricity demand. Unfortunately this is not the case. During windy periods all wind generators would produce electricity at the same time. The electricity production of larger wind farms may exceed the regional electricity

(*) E-mail: Georg.erdmann@tu-berlin.de

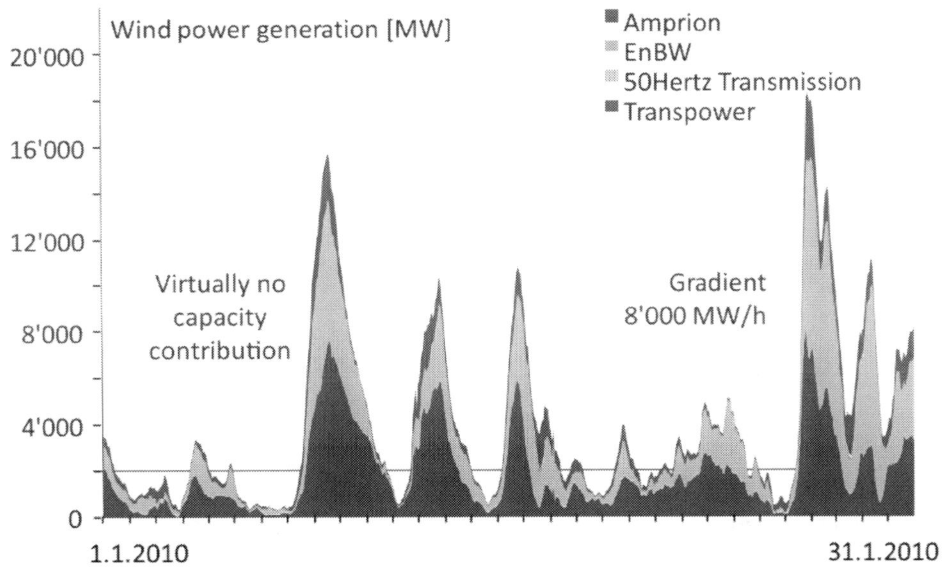

Fig. 1. – Volatility of wind power in Germany. Source: [1], p. 100.

demand. On the other hand during calm periods wind power generation may not be sufficient to cover the regional electricity demand.

An empirical example of the volatility of wind-based electricity generation is shown in fig. 1. It presents the feed-in of wind power into the grids of the four German regulation zones. The gradient of wind availability can be rather high. This requires conventional power generations to quickly increase or decrease generation. On the other hand even a strong increase of wind capacities will not prevent wind power generation deficits during calm periods of several days or weeks. The German wind power generation is still not sufficient to exceed national electricity demand, but with growing wind capacities the number of hours with excess generation will increase. This will increase the problems associated with the volatile nature of wind power generation.

In winter months solar capacities are hardly able to fill this gap. Daylight periods are short and solar insulation weak. But during the summer period solar energy, in particular photovoltaic, can contribute to the aggregated electricity demand, but with the same volatility problems as wind energy, for example excess generation in sunny hours around noon and no generation during the night.

From an engineering point of view, several concepts exist to solve the described intermittency problems:

- Curtailment of wind power generation if necessary.
- Providing dispatchable and flexible backup generation capacities such as geothermal or hydropower plants, biogas plants, and also certain fossil-fired power plants.

DOI: 10.1051/epjconf/201714800005

– Strengthening the transmission grid capacities over long distances in order to balance electricity supply and demand over larger geographical regions.

– Introducing demand side management with the intention to provide the balance between electricity demand and generation through flexible electricity customers.

– Large-scale storage capacities with pumped hydropower stations, batteries or electrolysis capacities which transform excess electricity into hydrogen or other chemical energy fuels. The stored energy can be used for power generation during deficit periods.

The engineering concepts, technologies additional costs of these options vary significantly. An important issue is that the costs per energy unit provided depend not only on the unit expenditures of the associated investment, but also on the number of hours per year during which the capacities are to be used. As example, the capital expenditures per unit of electricity delivered are ten times as large if the equipment is used during only 500 hours per year instead of 5000 hours per year. Section **2** looks at this aspect more in detail.

A follow up question is how the annual utilization period should be determined. If the dispatch of electricity technologies is determined by a central planner, the answer is trivial. More complex is the answer if the dispatch is based on the laws of market economics. In this case the annual utilization hours of each technology depend on its position in the so called merit order. The merit order concept explains both the annual utilization periods of individual technologies and the volatility of wholesale electricity prices. It is therefore quite central for understanding electricity economics and will be explained in sect. **3**

Based on the merit order concept, the remaining sections of this paper discuss market entry barriers of new technologies which may be needed for integrating volatile renewables into the electricity system. They also present ideas how to overcome these barriers.

2. – Economics of renewable-energy investment

Any economic analysis of investment decisions requires forecasting the cost and revenue streams over the expected lifetime of the planned investment. But only those economic variables have to be considered that are affected by the investment decision. If a company plans for an incremental expansion of capacity, the capital expenditures of the given installation as well as the management costs should be excluded, for example. However, the additional expected sales revenues per period generated by the investment are relevant. In the electric industry they depend on:

– the capacity Cap to be installed (measured in megawatt electricity generation),

– the capacity factor ν, specifying the expected percentage of annual full-load operation. As the year has $24 \cdot 365 = 8\,760$ hours, a capacity factor of $\nu = 20\%$ equals $0.2 \cdot 8\,760 = 1\,752$ full-load operation hours,

– the average expected price of electricity p_E (wholesale electricity price).

Given the power plant capacity Cap (in MW), the expected annual output Q and the corresponding annual revenues are:

(1) $$Q = Cap \cdot \nu \cdot 8\,760 \text{ (in MWh)},$$

(2) $$p_E \cdot Q = p_E \cdot (Cap \cdot \nu \cdot 8\,760) \text{ (in Euro/MWh)}.$$

The financial counterparts of the annual revenues are the expected annual costs. They can be divided into variable costs and fixed costs. Variable costs C_{var} include costs for intermediate inputs such as fuels, emission rights, waste disposal, and to some extend wages (given flexible employment contracts). Variable costs per unit output $c_{var} = C_{var}/Q$ are obtained by dividing expected variable costs through the expected output Q.

The fixed costs are the annualized costs of the investment. A simplified analysis assumes the investment expenditure to take place in period t_0, leading to a negative cash-flow Inv_0 in this period. In the following years, the cash flows are $(p_{Et} - c_{var,t}) \cdot Q_t$. They should usually be positive in order to make the project economically viable. The time horizon is the end of economic project life T. A time series of annual cash flows can only be meaningfully evaluated if the individual cash flows are referenced to one period, for example $t = 0$. This is done by calculating the Net Present Value

(3) $$NPV = -Inv_0 + \sum_{t=1}^{T} \frac{(p_{E,t} - c_{var,t}) \cdot Q_t}{(1+i)^t}.$$

In this formula, all future cash flows are discounted with the interest rate i. The term $(1+i)^{-t}$ is referred to as the discount factor. The economic evaluation of an investment project is positive if $NPV > 0$.

If the annual cash flows are to remain constant over the entire project life, eq. (3) can be rewritten to

(4) $$NPV = -Inv_0 + (p_E - c_{var}) \cdot Q \cdot \sum_{t=1}^{T} \frac{1}{(1+i)^t}.$$

The sum on the right-hand side is the Present Value Factor of an Annuity

(5) $$PVF_{i,T} = \sum_{t=1}^{T} \frac{1}{(1+i)^t} = \frac{1}{i} - \frac{1}{i \cdot (1+i)^T}.$$

In order to calculate the break even conditions, one has to set $NPV = 0$ in eq. (4). By solving for the electricity price p_E, one obtains is the break-even price that is required

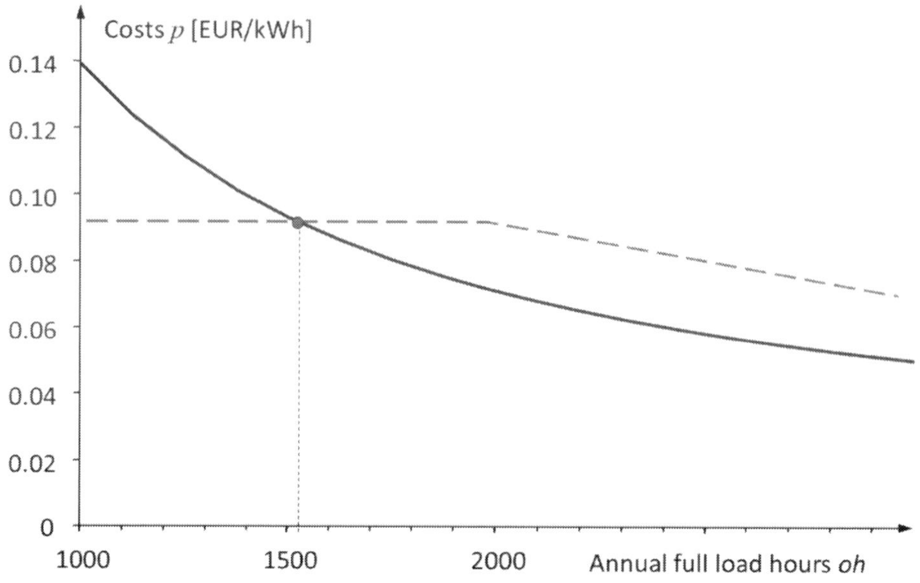

Fig. 2. – Example for the LCOE of wind power.

to cover the investment outlay Inv_0 and the unit variable cost c_{var}:

$$p_E = \frac{Inv_0}{Q \cdot PVF_{i,T}} + c_{var} = LCOE. \quad (6)$$

This is the levelized cost of electricity $LCOE$. Figure 2 presents the $LCOE$ of wind power as a function over the annual full-load operation hours (continuous line). The parameters assumed for this example are

$INV_0 = 975$, EUR/kW unit investment costs,

$c_{var} = 42$, EUR/kW unit operation costs (1–10 year),
58, EUR/kW unit operation costs (11–20 year),

$i = 5$ percent interest rate,

$T = 20$ years of economic use.

Figure 2 also shows the level of the German support for wind electricity (fixed feed-in premium, dotted line). Comparing this support with the $LCOE$, the conclusion is that wind turbines would allow an economic operation if they are installed at locations where the wind conditions would allow at least 1500 full load hours (break-even location). Investing at locations with better wind conditions would lead to significant profits, while wind installations at bad wind locations would be uneconomic. The example in fig. 2

presents a range of 0.05 to more than 0.14 Euro/MWh for the *LCOE* of wind power production.

The strong dependency of *LCOE* on the annual utilization hours is relevant for all types of electricity technologies: power generation plants, storage capacities, grid investments, combined heat and power installations etc. It is also relevant for demand side technologies such as electrolyzes, electrical heat pumps, etc. As a rule of thumb, capital intensive technologies(¹) require a large number of annual utilization hours, for example more than 5000 hours per year. If the market conditions do not allow such large numbers, the technologies are typically non-economic unless some form of state aid is granted. The emancipation from such political interventions requires efforts towards reducing the specific investment expenditures, in particular if annual utilization hours are physically limited.

3. – Merit order effect of renewable electricity

3˙1. *Power plant dispatch on day-ahead markets.* – This section explains the dispatch of electricity technologies according to market rules based on [2]. Starting point is the idea that on perfectly competitive markets each existing generation unit reaches its profit maximum if the dispatch is organized according to its marginal costs. This is shown by simple mathematics: It is assumed that the operator tries to maximize the periodic profit

$$\Pi(Q) = \bar{p}_E \cdot Q - C(Q) \tag{7}$$

(difference between the periodic revenues $\bar{p}_E \cdot Q$ and the periodic costs $C(Q)$). The optimal solution is ound by setting the derivative of the profit function with respect to the produced quantity Q equal to zero:

$$\frac{\mathrm{d}\Pi}{\mathrm{d}Q} = \frac{\mathrm{d}(p_E \cdot Q)}{\mathrm{d}Q} - \frac{\mathrm{d}C}{\mathrm{d}Q} = \bar{p}_E - \frac{\mathrm{d}C}{\mathrm{d}Q} = 0 \quad \rightarrow \quad \frac{\mathrm{d}C}{\mathrm{d}Q} = \bar{p}_E. \tag{8}$$

If the operator has no market power (or is not allowed to exercise market power), it does not influence the sales price p_E so that this price is exogenous: $p_E = \bar{p}_E$. The operator should adjust the individual production Q so that the marginal cost $\mathrm{d}C/\mathrm{d}Q$ —the additional expenditures associated with the generation of one additional electricity unit [MWh]— is equal to the fixed sales price $\bar{p}_E$(²). If the marginal cost is smaller, the supply should be expanded, otherwise it should be reduced. Any sale at a price below the marginal costs is economically stupid.

Depending on the type of plant, marginal costs are the sum of several cost components:

- Additional costs for fuel needed to generate one unit of electricity. It depends, among others, on the fuel efficiency of the plant. It is worth to note that for

(¹) These are technologies with a large ratio "fixed cost/variable cost".
(²) If the cost function $C(Q)$ is linear, the marginal cost is equal to the variable unit cost c_{var}.

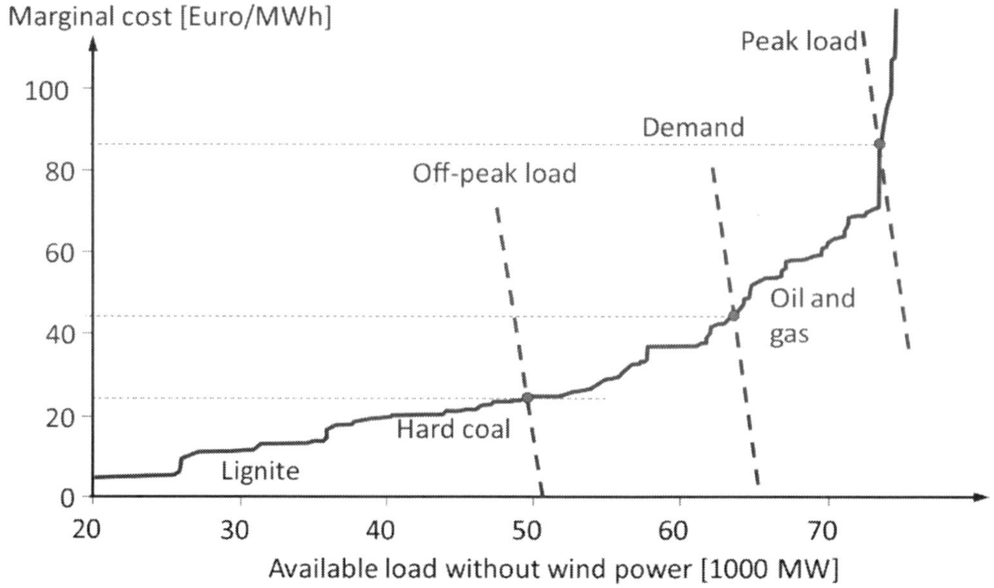

Fig. 3. – Merit order of German power plants.

wind and solar generators these costs do not exist so that their marginal costs are typically lower than the marginal costs of any conventional power plant.

– Additional costs for emission rights (if applicable). The corresponding contribution to the marginal cost depends on the tons of CO_2 emitted per MWh of electricity, multiplied by the CO_2 market price. Again, for wind and solar generators this cost component does not exist.

– If the output of thermal power plants falls below the optimal design level, efficiency losses occur and must be taken into consideration when calculating marginal costs.

– Fuel losses during start-up and shut-down periods.

– Accelerated wear of the equipment (boilers and pipes) due to thermal stress resulting from temperature change.

The investment costs and other costs such as personnel or administration expenditures are not affected by the generation decision and are therefore irrelevant for the evaluation of such decisions.

The aggregate electricity supply curve results from ordering the available electricity supply capacities of a country according their marginal costs. This is the merit order. Figure 3 shows a typical merit order for Germany (solid line). Compared to fossil fuel plants, wind and photovoltaic generators have rather low marginal costs and therefore stand at the beginning of the merit order. Nuclear and lignite power plants come next, whereby new lignite plants are before old plants due to their advanced fuel efficiency. Next in the merit order are hard coal plants, followed by Combined Cycle Gas Turbines (CCGT), open cycle gas turbines and older oil and gas plants with comparatively low fuel efficiencies.

Figure 3 also shows the electricity demand for three individual hours. The intersections with the aggregate supply curve (or the merit order curve) give the market clearing conditions for these hours, the produced electricity and the Market Clearing Price (MCP). The ask price of the last operator that is accepted —the last capacity that is dispatched— sets the MCP of the particular hour, while all other dispatched generators earn a contribution margin on top of their marginal costs which allows to cover (some of their) fixed costs. Due to the convex pattern of the merit order curve, hourly power prices for peak-load hours are higher and more volatile than hourly prices in off-peak periods.

The hourly MCP can be observed on so called day-ahead markets. The most liquid day-ahead market in continental Europe is organized by the European Power Exchange (EPEX). It is an auction market: All bids and asks for individual hours are aggregated into an hourly supply and a demand curve (see table II). After gate closure, the EPEX calculates the hourly MCP and informs market participants and the public about the trading results. Generators who have asked a price above the MCP are excluded from selling their generation on the day-ahead market. They have to curtail their generation accordingly as far as they cannot find a buyer for their generation on intraday markets. In this way, the number of annual operation hours is determined on free and competitive electricity markets.

According to fig. 3 more power capacity is dispatched in peak load periods than in off-peak periods. If suppliers offer electricity at their individual marginal costs (what they should do in order to maximize profits or minimize losses), the cheapest portfolio of capacities —in terms of marginal costs— is dispatched. Therefore capacities with low marginal costs achieve much more annual utilization hours than capacities with high marginal costs.

3·2. *Impacts of renewables on the merit order.* – Having almost no marginal cost, wind and photovoltaic capacities stand at the beginning of the merit order. On free electricity markets they are dispatched whenever they are physically available, but their capacities may be curtailed if their generation —together with must-run capacities (for example Combined Heat and Power CHP)— exceeds the actual electricity demand.

The merit order curve moves horizontally according to the volatile availability of renewable electricity. If more wind and photovoltaic electricity is available, the merit order curve moves to the right. This has two implications:

- Given a certain quantitative demand, other generation capacities have to interrupt their power generation during periods with high availability of wind and photovoltaics. This refers to capacities with relative high marginal costs, which are actually gas-fired power stations, in particularly combined cycle gas turbines (CCGT). Coal-fired power stations are less concerned as their marginal costs are actually

TABLE I. – *Merit order effect estimates of wind and PV in Germany, 2006–2012.*

	2006	2007	2008	2009	2010	2011	2012
				Euros/MWh			
Sensfuß *et al.* (2008) [3]	−7.8						
Weigt (2009) [4]	−6.2	−10.4	−13.0				
vbw (2011) [5]					−8.0		
Sensfuß (2013) [6]		−5.8	−5.8	−6.0	−5.3	−8.7	−8.9
Speth, Stark (2012) [7]					−5.6	−5.6	
Cludius *et al.* (2013) [8]			−10.8	−7.8	−6.0	−7.7	−10.1

below those of relative climate friendly gas power plants([3]). With the growing share of renewables, coal-fired power stations will also have to interrupt generation during power surplus periods.

– Second, as wind and photovoltaic generation squeezes out relative expansive gas-fired power generation, the Market Clearing Price (MCP) on the day-ahead electricity market becomes smaller. This effect is called the "merit order effect of renewables" (see [3] and [6]). Due to the merit order effect of renewables, the wholesale electricity prices have fallen below marginal cost even of highly energy efficient CCGT. Such investments are no more economic today.

Several studies have quantified the merit order effect of renewables. An overview for Germany is presented in table I. Most of the studies cited here are based on a structural power market model based on a large databank of several hundred generation units. The hourly merit order function is calculated from the marginal costs of the capacities available in the respective hour. By comparing the merit order function with the hourly electricity demand, the Market Clearing Price (MCP) results. For estimate the merit order effect of renewables, two scenarios are compared: one using the historical renewable power generation and one without any renewable generation. Comparing the yearly average of day-ahead price between these two model runs gives an estimate for the merit order effect of renewables.

There are some challenges associated with structural market models: Usually the scenario calculated with the renewable power generation should reproduce the realized hourly day-ahead prices, but this is not automatically the case because real behavior of power plant operators may differ from the implicit assumption of cost optimal dispatch of generation capacities. In addition, some capacities are not available for day-ahead trading. Finally, public information about individual costs for fuels is not public. Therefore structural market models require some calibration before they are able to reproduce market results.

([3]) This is because gas prices are high compared to coal prices and because CO_2 prices are presently rather low.

DOI: 10.1051/epjconf/201714800005

Another approach for calculating the merit order effect of renewables is used in [8] and [5] and based on an econometric model (reduced form model). Starting point is a time series of hourly wholesale power prices (endogenous variable) which is assumed to be statistically correlated with exogenous variables such as fuel and CO_2 prices, hour-of-day effects, calendar effects (weekends, holidays, country wide vacations) wind and photovoltaic generation etc. The parameters describing the impact of the exogenous variables on the endogenous variable are calculated by an econometric estimation. Applying the model, two cases are simulated and compared, one with and one without wind and photovoltaic power generation. The second case results from simply setting the historic wind and photovoltaic generation to zero.

The merit order effect of renewables calculated with econometric approaches is in the range of results achieved by structural market models. Thus the quantification of the merit order effect turns out to be quite robust.

It can be assumed that the merit order effect of renewables has grown since 2012, because the renewable generation capacities in Germany are 20% larger in 2014 than in 2012. To check the consequences, [9] presents an updated econometric estimation of the merit order effect of wind and photovoltaics in Germany, based on the three year period between 2012 and 2014 (with the range of 26304 hours). According to this study, the merit order effect has reached 12 Euro/MWh in 2014. To put this result into perspective, it states that without electricity from wind and photovoltaics the day-ahead price 2014 in Germany would have been 44.80 Euro/MWh instead of 32.80 Euro/MWh. The downward price trend is continuing; in 2015 the average day-ahead price declined further to 31.60 Euro/MWh and has fallen below 25 Euro/MWh in the first half of 2016. Obviously this is a dramatic development for all those who produce and sell electricity on the wholesale market.

In the European single market restrictions to trade are prohibited. Therefore the German electricity price collapse spreads to neighboring countries. Power purchases from the attractive German wholesale market become attractive for foreign market participants. As a consequence, German net electricity exports have passed 8% of national generation. This puts a pressure on foreign wholesale prices are forces power generators in neighboring countries to curtail production([4]). In sum, the renewable capacity growth in Germany challenges not only the domestic electricity generation but also the generation in neighboring countries, in particular gas-fired power plants with their relative high marginal costs.

4. – Intermittency of renewables: Do we need capacity markets?

With the growing generation from renewable sources, conventional generators are exposed to twofold losses:

– reduced annual operation hours,

([4]) For example, the wholesale electricity price in Italy would be 2.5 Euro/MWh higher without the German merit order effect of renewables, according to our estimates.

 DOI: 10.1051/epjconf/201714800005

– reduced revenues from electricity sales.

Thus the economic survival of conventional power generation is at risk, implying that existing plants are prematurely taken out of order and new investments into conventional generation are withdrawn. Different to conventional generators, state aid in favor of renewable generation makes investments into wind and photovoltaic still economical so that such investments are going on. But through the volatile nature of their generation, these capacities alone are not able to secure electricity demand.

Any generation system with rather high shares of wind and solar technologies needs some backup capacities, based on a combination of

– long-distance power transmission,

– large-scale electricity storage capacities,

– demand side management.

But all these options are costly and/or faced with other problems and bottlenecks. Therefore investments into dispatchable generation capacities are assumed to be inevitable.

Electric industry has much experience how to secure supply. As far as electricity cannot be stored, a reliable power system requires some excess and reserve generation capacities. The excess capacities are needed to cover unexpected loads, while reserve capacities are required to back-up scheduled and unscheduled generation outages and/or power demand spikes. In addition, some flexible generation capacity is required for supplying regulation power and other grid services, such as redispatch of generation in case of local bottlenecks in the transmission grid. As a consequence, reserve margins of about 10 percent of the maximum load are assumed to be necessary to secure electricity supply at any time (system adequacy).

But how this system adequacy can be reached? If the electric generation is dominated by one monopolistic utility, the solution is easy: The regulator requires the utility to guarantee the needed generation capacities according to a mandatory system adequacy level and allows the utility to recover the costs through electricity tariffs. But on competitive markets such an automatism is not feasible because supply security has the character of a public good: Each competitor is interested in avoiding the associated costs for securing electricity because this would allow him a competitive advantage on the power market.

Knowing this dilemma, the electricity market regulators have developed mechanisms in order towards securing electricity supply (based on the pioneering work of [10]):

– The (still) monopolistic Transmission System Operator (TSO) is requested to purchase capacities for supplying regulation power. The sellers of regulation power are determined by regularly pay-as-bid auctions and receive capacity payments if awarded. In return they must activate the awarded capacities at the disposal of the TSO.

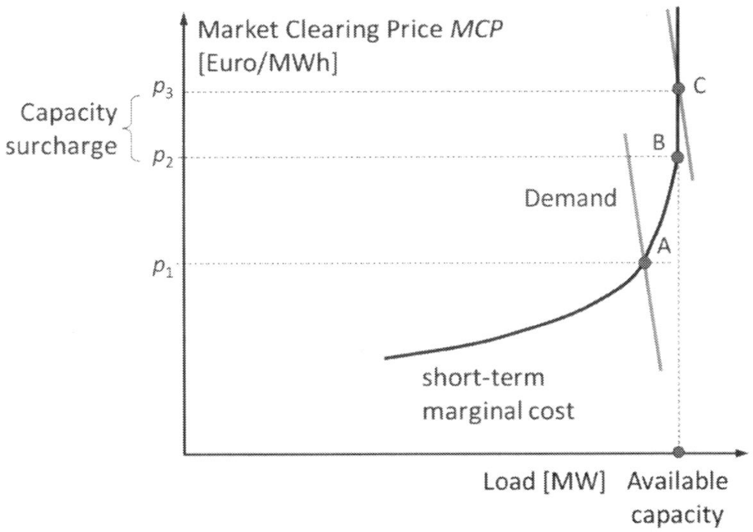

Fig. 4. – Electricity market under capacity shortage.

- In a liberalized market with retail competition each wholesale market participant (balancing group managers) has to purchase balancing power from the TSO to adjust for imbalances between supply and demand in its balancing group. If the imbalances are excessive, the regulator may impose fines and other sanctions. In extreme situations the TSO can even exclude balancing group managers from using the electric grid.

- In addition backup power markets exist where option contracts for compensating eventual plant outages are traded. The demand comes, among others, from the operators with small generation portfolios. Large generators can hedge the outage risks by own generation reserves.

Under the label "capacity market", a controversial discussion on additional instruments for securing electricity demand is presently underway (for example [11] and [12]). To explain this debate, fig. 4 is presented showing a simplified merit order function (similar to the concept explained in [13], p. 70f). Point A corresponds to a regular market situation in which the marginal cost of the last available power plant (*i.e.* excluding the reserve margin) sets the Market Clearing Price p_1. Because the day-ahead market is organized as a unified price auction, all power plants to the left of this last plant sell their electricity against the same price p_1 and thus earn (infra-marginal) rents on top of their marginal costs.

Point B corresponds to the last available capacity unit in the merit order with marginal cost p_2. In hours with regular market conditions the Market Clearing Price is below p_2 because available capacities are sufficient to cover the hourly demand. But in situations with high power demand and low wind and solar generation the aggregated available capacities may be insufficient. Point C in fig. 4 explains this case: There is no intersection between the demand and the merit order. But if the merit order curve is extended vertically above point B, an intersection in point C exists. The resulting MCP is $p_3 > p_2$ defining a positive price margin $p_3 - p_2$ which is called "capacity surcharge" or "capacity rent". The MCP exceeds the marginal cost of any generation capacity in the merit order. If such margins are observed, traditional competition theory would assume misuse of market power. But this is not the case. The price spike to p_3 is simply a market reaction to solve a capacity shortage problem and is necessary for balancing current discrepancies between (low) electricity supply and (high) electricity demand. If such situations are expected to occur frequently, the likelihood of premature closures of existing generation capacities declines and that of investments into new generation capacities increases, thereby solving the system adequacy problem.

There is some discussion among economists whether unpredictable occasional price spikes are sufficient for securing investments into power generation. From the investor's point of view such price spikes are basically price risks and need to be compensated by higher average wholesale prices. This can be seen as the disadvantage of markets solving the system adequacy problem.

But the market approach has other problems: Volatile and spiky wholesale power prices can be rather unpopular so that electricity customers may organize some political pressure in favor of lowering wholesale power prices. This pressure is typically orchestrated by the public reproach that generators manipulate market prices for obtaining excessive profits on behalf of their customers. If regulators follow these claims, they impose some explicit or implicit upper limit on wholesale power prices. In this case the wholesale power market cannot raise sufficient revenues for financing power plant investments. In the literature this is discussed as the "missing money problem" (see for example [14]).

The missing money problem may be solved by introducing capacity payments to certain power generators. If governments do not want to spend taxpayer's money, the capacity payments shall be raised from final electricity customers in form of capacity levies (similar to renewable-electricity levies). The selection of the plants may again be organized by auctions. In this case the regulator defines the auction frequency, the aggregated auction volumes and the conditions that successful plant operators must satisfy. Awarded power plants are taken out of the regular electricity dispatch and are scheduled by request of the regulator. This mechanism will increase the wholesale power prices during regular periods, but reduces the wholesale price in emergency situations, according to the criterion for releasing emergency capacities.

There is a "decentralized" alternative to a regulated auction based capacity market: Power generators and retailers are required to always keep their electricity schedules balanced. In order to stick to this rule, retailers may purchase an equivalent number of capacity guarantees issued by flexible generators or flexible demand. The same holds for power generators, in particular if they feed electricity from volatile sources into the grid. Flexible generators, electricity storage operators and final electricity customers with interruptible loads are typically on the supply side of capacity guarantees. The price for these guarantees shall be high in periods with tight capacities (weak renewable-electricity supply), but it can also be zero if sufficient capacity is available (strong renewable-electricity supply). The beauty of this "decentralized capacity market" is that its implementation can be left to the market. Innovations and competition will quite likely lead to cost-efficient solutions. The regulator can limit itself to enforce the balancing through sanctions on market participants that do not comply.

In addition to regulated and decentralized capacity markets a third concept may be used to solve system adequacy problem. It is based on the customer's choice to select the desired quality of electricity supply. Retailers may offer power supply contracts that give them the right to cut the supply during some hours per day, if capacities are short and prices high (usually during evening hours). Electricity customers can opt for a non-interruptible power supply but they have to pay a surcharge according to the individual cost associated with this service. The acceptance of interruptible supply contracts depends on the customer's value of lost load (VOLL; defined in [13], p.154ff, see also [15]) which varies according to time, power application, availability of own generation, electricity storage capacities etc. Such a concept would again initialize a variety of innovations so that the final result will be cost efficient compared to centrally planned capacity markets or mandatory investments into generation (or electricity storage) capacities.

5. – Using volatile excess electricity: "Sector coupling"

To conclude the former analysis, a manifold of possible approaches exists towards securing electricity supply in power systems with rather high shares of volatile generation capacities. The other side of the medal is that in such systems there will be many hours with excess generation.

Figure 5 presents such a constellation. The figure shows the cumulated electricity demand (negatively sloped bid curve) and supply (positively sloped ask curve) at the European Power Exchange EPEX for a Sunday afternoon in May 2016. In this hour the Market Clearing Price (MCP) was strongly negative, namely −130.09 Euro/MWh. This means that generators had to pay for selling electricity and buyers had earned money for purchasing electricity for this hour.

This constellation is characterized by low electricity demand and high electricity supply. But one has to ask why generators do not cut production, instead of paying for selling electricity. Several arguments can be brought forward for this behavior:

- (Renewable) Generators that would lose state aid in case of curtailing generation.
- Thermal power stations with technically limited generation flexibility.
- Must run capacities, for example generators that are obliged to provide regulation power.

For these reasons, wholesale power prices can become negative from time to time.

Even in periods when power prices are positive but close to zero, an economic incentive exists for final users to purchase more electricity. This additional demand would be helpful for integrating renewable electricity into electricity markets. If the additional electricity demand is used to replace fossil fuels in other energy sectors, it would also

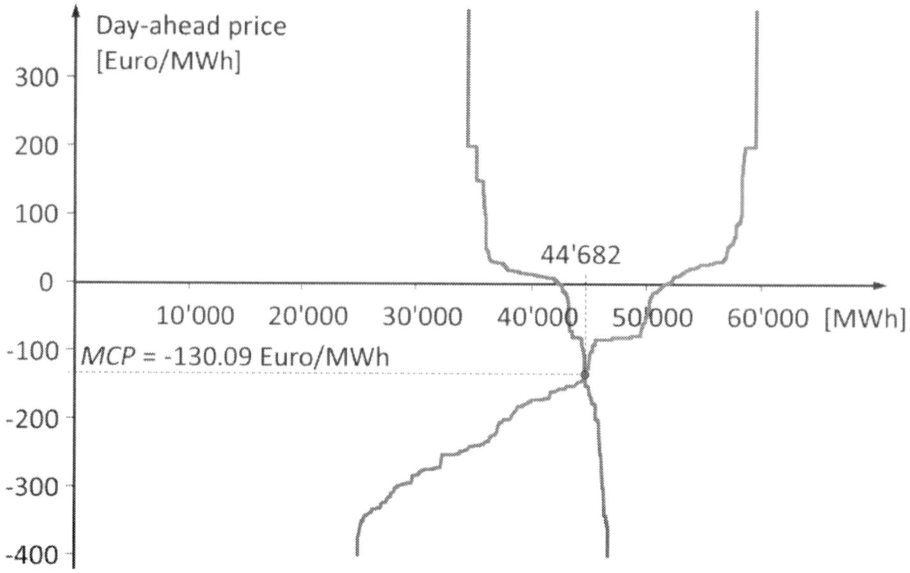

Fig. 5. – Bid and ask curves on the German day-ahead market for the hour 14–15 h on 8 May 2016.

contribute to the reduction of greenhouse gas emissions. The use of electricity in other sectors for this purpose is also known as "sector coupling" (see [16]).

Sector coupling examples are power-to-heat appliances in combination with hot water storage capacities. Heat production from electricity is usually regarded as energetically unattractive due to the associated exergy losses, but regarding the electricity system as a whole, the losses would be larger if the available renewably electricity would be curtailed due to insufficient power demand. In many cases relative minor investments are needed for developing power-to-heat capacities, because most central heating installations as well as some district heating systems are equipped with heat storage capacities. The number of hours with low or negative wholesale power prices will be small (at least for the time being), but regarding the low specific investments of power-to-heat systems, their economics should be attractive in principle even if their annual operation will not exceed 1000 hours per year([5]).

But for final electricity users power-to-heat systems are only economical if the resulting heat costs are below the final user prices of natural gas or heating oil which are in the range of 30–70 Euro/MWh. No problem would exist if the electricity price for final users is equivalent the wholesale electricity price: Power-to-heat makes only sense in periods with excess electricity from renewables, and just these periods are characterized by rather low and eventually negative wholesale prices.

([5]) See sect. **2**, in particular fig. 2, which highlights the basic mechanism between annual operation hours and levelized costs of energy. This mechanism applies not only for wind energy but also for power-to-heat systems and other electricity technologies.

TABLE II. – *Electricity price components in Germany in the year 2016.*

	Power purchase from the grid	Auto generation	Generation in „local context"
		Euro/MWh	
Grid fee	18.00 – 30.00		18.00 – 30.00
REN levy	63.54	22.24	0.00
Electricity tax	20.50	Legally unclear	
Concession fee	1.1 – 23.90		1.1 – 23.90
CHP levy	4.45		4.45
§ 19 StromNEV	3.78		3.78
Offshore levy	0.40		0.40
Total	86.77 – 121.57	22.24 – 42.74	23.73 – 83.03

Unfortunately the electricity prices for final users are by far higher than the electricity wholesale price. It includes costs of retailers (for marketing, purchasing, scheduling and metering), grid fees, renewable-electricity levies, electricity taxes. Table II presents an overview of the present situation in Germany. The situation in Germany is rather complex because of many special circumstances defined in the federal law. But as a general conclusion, the retail prices of natural gas are hardly met by retail electricity prices except in singular cases. This challenges the market introduction of power-to-heat systems.

Power-to-heat-systems may have only a chance if the price for electricity used for running these systems is sufficiently small. There are basically two approaches to get there:

- Define specific power-to-heat applications for which (most of the) grid fees, levies, fines and taxes have not to be paid. This approach is problematic because the regulator will hardly be neutral with respect to different concepts, technologies and investors. As a result, the success of power-to-heat does not depend on the specific competitiveness but on the specifications defined by the regulator.

- Shift the reference for grid fees, levies and perhaps taxes from energy (kWh purchased) to load (maximum electricity load in kW during a calendar year). As a consequence, the additional cost of electricity used for power-to-heat systems can become rather low, as far as the operation of these systems keeps the annual maximum load unaffected.

As a conclusion, it should be possible to modify the electricity tariff structure in such a way that power-to-heat systems may generally become economical for final users so that excess electricity from volatile renewable sources will be used on a large scale for substituting fossil fuels in the heat market.

Other sector coupling options are based on power-to-gas and power-to-liquid concepts. The core of such systems is the electrolysis of hydrogen. Subsequent processes may use this hydrogen to produce synthetic methane or methanol. There are two shortcomings here. One is that the energy losses of these processes are significant. The second is that the specific investment expenditures are relatively high, today at least 3000 Euro/kW installed capacity. Therefore the levelized costs become rather high if the operation of these systems is restricted to periods with excess electricity from renewable sources (see again table II). The operators have an economic interest to run these systems in base load, but this would be basically incompatible with the aim of integrating volatile renewable sources into the electricity market.

6. – Economics of electricity storage

A final sector coupling option is batteries. Due to the market growth for battery electric vehicles their costs are declining strongly. Therefore it appears likely that electricity will soon replace some fossil fuels in the transportation sector. But this process will not only be controlled by battery costs but also by the driving ranges of a fully charged battery vehicles and the achievable number of battery charging cycles (the economic lifetime of batteries in vehicles). The deeper discussion of these topics goes beyond the scope of this paper. This paper rather addresses the stationary utilization of batteries (and other electricity storage devices, for example hydropower pumped storage capacities) and explains some particularities concerning the economics of energy storage.

Independent of their physical and technical concept, the operation of electricity storage devices is characterized by creating additional demand for electricity in period of excess supply and additional supply in periods of electricity shortage. If dispatched in this way, storage technologies contribute to the market integration of volatile electricity generation.

The proper dispatch of storage capacities has to focus on the wholesale electricity prices: The charging of any electricity storage should happen in periods of low wholesale electricity prices, while the discharge should take place in periods of high prices([6]). Thereby storage capacities contribute to the stabilization of wholesale electricity prices. The revenues associated with this operation depend on the price spread between electricity for charging and electricity from discharging.

This price spread increase, among others, with the cumulated size of renewable-electricity capacities and shrinks with the mandatory introduction of capacity markets. The cumulated share of electricity storage capacities has also an impact. This is ex-

([6]) Insufficient grid capacities can also be a reason for curtailing renewable-electricity generation. A reliable concept for running electricity storage capacities is based on the elimination of such regional imbalances. Here the aim is so stabilize local distribution grids, independent of the current wholesale prices.

DOI: 10.1051/epjconf/201714800005

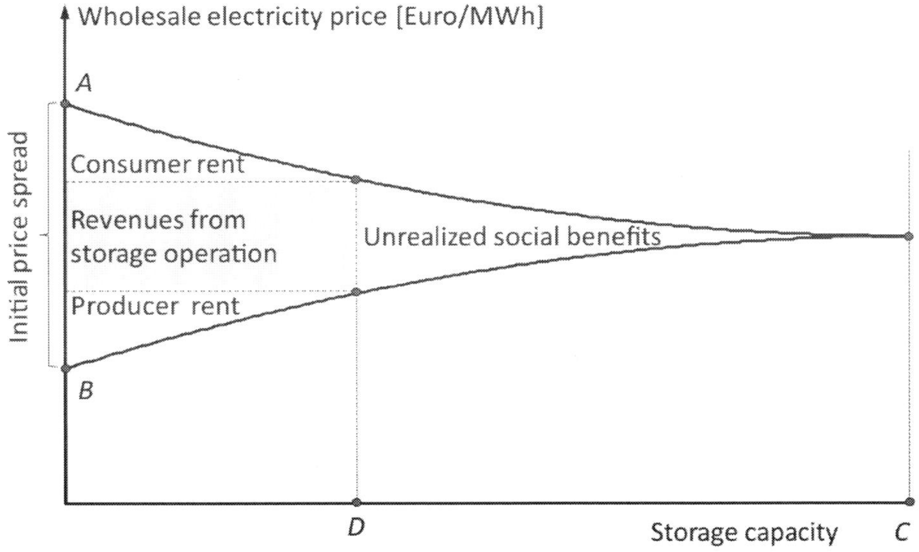

Fig. 6. – Cumulated storage capacity and power price spread (source: adapted from [1], p. 119).

plained in fig. 6. Without storage capacities the spread between the highest and the lowest hourly day-ahead price during a given period (24 hours of the day, for example) corresponds to the difference between the points A and B. The economic dispatch of electricity storage would reduce this spread, and there is a maximum storage capacity C which would imply the fully distinction of the price spread.

The revenues from an optimal operation of electricity storages correspond to the price spread which remains after the dispatch of the storage capacity, times the available storage capacity. Figure 6 presents an example. The rectangle corresponds to the cumulated storage revenues given the storage capacity D. The two triangles marked with "consumer rent" and "producer rent" indicate the social benefits associated with the reduced price spread resulting from the optimal operation of the storage capacities. The triangle to the right of the rectangle represents the unrealized social benefits due to the limited storage capacity.

Even if there would be no costs associated with investing and operating electricity storage capacities, the free market solution would not lead to the maximum storage capacity C because in this situation storage capacities would receive no revenues at all. Even if storage costs would be zero, the optimal storage capacity must be smaller than C. Of course storage costs exist and play a role. For the economic optimum it is crucial that the revenues exceed the investment and operation cost of the storage capacity. If this is not the case, no private storage investment is to be expected. For the social optimum the unrealized social benefits must be equal to the storage investment and operation cost. Thus a discrepancy between the private and the social optimum exists.

 DOI: 10.1051/epjconf/201714800005

Another conclusion from fig. 6 is that the insider-outsider problem is highly virulent: Owners of electricity storage capacities have an interest to hinder newcomers to enter the market, even if the economic storage optimum is not yet reached, because each newcomer reduces the price spread and thus the revenues of the insiders. This problem is relevant, whenever a business concept is based on a price spread. It is particularly relevant for electricity storage capacities because even small additions to the aggregated capacities lead to a strong decline of an initial high price spread. For understanding this statement the reader may consult fig. 5 again. It shows a situation with a strongly negative Market Clearing Price. But this price would have be significantly higher if there would have been some more MWh electricity demand (red curve shifts to the right) or some MWh lesser electricity supply (blue curve shifts to the left).

7. – Outlook

Wind and photovoltaic generation technologies have moved from the innovation phase to the phase of market application and diffusion. If invested at favorable wind or solar locations, the levelized cost had fallen below the levelized cost of new gas and coal power stations. But for the comparative competitiveness of these renewable-electricity sources the levelized costs are not the only determinant. In electricity systems there must be a balance between generation and demand in all periods. Electricity potentials from wind and photovoltaics are principally abundant but electricity generation from these sources is by nature rather volatile. This poses additional challenges that must be solved before wind and photovoltaic can achieve larger shares in future power systems.

This paper does not address physical and engineering properties and features of technologies that can be used for integrating renewables into electricity markets. It rather presents ideas about how to assess their economics. Three cases are distinguished:

– Approaches to address situations with lacking electricity generation.

– Approaches addressing situations with excess electricity generation.

– Approaches to address both problems through electricity storages.

The paper concludes that no fundamental reasons exist why large market shares of renewables are not feasible, but additional system costs in one or another way are inevitable and the question arises how to cover these costs. This paper proposes some concepts, but without claiming to present a comprehensive assessment of all possible models and solutions which could come to one's mind. Regarding the dimension of the challenges and costs, the paper prefers solutions in which governments restrain from technology specific interventions but rather leave the selection of technologies to fair competition and market forces. Experience shows that this would lead to better and cost-efficient solutions. But governments have a role to play: They must create an appropriate environment for competition among technologies. According to the paper, here is the key deficit for integrating volatile generation into electricity markets which is needed in order to further increase the share of renewable electricity.

REFERENCES

[1] Ehlers N., *Strommarktdesign angesichts des Ausbaus fluktuierender Stromerzeugung. Reichen die Investitionsanreize für Backup-Kraftwerke und Stromspeicher?* (Dissertationen.de, Berlin) 2011.
[2] Boiteux M., *The Choice of Plant and Equipment for the Production of Electric Energy*, in *Marginal Cost Pricing in Practice*, edited by Nelson James (Prentice-Hall, Englewood Cliffs, NJ, 1964).
[3] Sensfuss F., Ragwitz M. and Genoese M., *Energy Policy*, **36** (2008) 3086.
[4] Weigt H., *Appl. Energy*, **86** (2009) 1857.
[5] vbw, *Kosten des Ausbaus der Erneuerbaren Energien* (vbw – Vereinigung der Bayerischen Wirtschaft e.V., München) 2001, www.baypapier.com/fileadmin/user_upload/Downloads/Standpunkte/Studie_Kosten_Erneuerbare_Energien.pdf.
[6] Sensfuss F., *Analysen zum Merit Order Effekt erneuerbarer Energien* (Fraunhofer-Institut für System- und Innovationsforschung (ISI), Karlsruhe) 2013, www.impres-projekt.de/impres-wAssets/docs/Merit Order-2012_final.pdf.
[7] Speth V. and Stark M., *The impact of wind and solar on peak and off-peak prices - Evidence from a two year price analysis*, paper presented at the *2012 IAEE European Conference, Venice, September 2012*, www.webmeets.com/files/papers/AIEE/2012/451/Abstract_IAEE_2012_Electricityprices.pdf.
[8] Cludius J., Hermann H. and Matthes F., *The Merit Order Effect of Wind and Photovoltaic Electricity Generation in Germany 2008–2012*, Working Paper 3-2013 (CEEM, Sydney, Australia) 2013.
[9] Praktiknjo A. and Erdmann G., *Energy J.*, **37** (2016) 89.
[10] Chao H. P. and Wilson R., *Resource Adequacy and Market Power Mitigation via Option Contracts* (Electric Power Research Institute (EPRI), Palo Alto) 2004.
[11] Joskow P. L. and Tirole J., *Rand J. Econ.*, **38** (2007) 68.
[12] Joskow P. L., *Util. Policy*, **16** (2008) 159.
[13] Stoft S., *Power System Economics: Designing Markets for Electricity* (Wiley-IEEE Press) 2002.
[14] Newbery D., *Missing Money and Missing Markets: Reliability, Capacity Auctions and Interconnectors*, Working Paper 1508 (EPRG, Cambridge, UK) 2015.
[15] Praktiknjo A., *Sicherheit der Elektrizitätsversorgung, Das Spannungsfeld von Wirtschaftlichkeit und Umweltverträglichkeitn* (2013).
[16] Mathiesen B. V. and Lund H., *IET Renew. Power Gener.*, **2** (2009) 190.

The road to sustainable transportation

Jo Hermans

Leiden University – P.O. Box 9504, 2300 RA Leiden, The Netherlands

Summary. — Since most modes of transportation rely on liquid energy carriers like gasoline or diesel, the approaching end of the fossil fuel era poses a special challenge. One should bear in mind that nothing matches the energy density of liquid fuels. It is, therefore, important to consider the efficiency of various modes of transportation and explore alternative options for energy carriers. This includes a feasibility study of the use of solar energy in transportation. In this perspective various modes of transportation are briefly reviewed and the possible alternative energy carriers assessed.

1. – Physics of transportation

In order to obtain insight into the possibilities of improving the performance of transport, it is useful to first turn to the fundamental physics of transportation. The key here is the *resistance* suffered by any vehicle moving at a certain speed, because the resistance directly determines the energy use per unit distance. In other words: the resistance in newtons equals the energy use in J/m, or —expressed more conveniently— in kJ/km.

For vehicles on wheels the resistance has two components: rolling resistance F_{r} and air resistance or aerodynamic drag F_{d}. For the latter contribution one should bear in

DOI: 10.1051/epjconf/201714800006

TABLE I. – *Typical values of the aerodynamic drag coefficient* C_d (*source for specified cars brands:* `https://en.wikipedia.org/wiki/Automobile_drag_coefficient`).

Flat surface	$C_d \approx 1$
Citroen 2CV	0.51
Clumsy 1960s car	0.4
Citroen C4	0.28
Toyota Prius	0.25
Tesla model S	0.24
Nuna solar car, World Solar Challenge winner	0.07
Ideal streamline	0.05

mind that the flow is not laminar but turbulent. So the total resistance can be written as

$$F = F_r + F_d = C_r \times mg + C_d \times A \times (1/2)\rho v^2, \quad (1)$$

where C_r is the rolling resistance coefficient, m is the mass, g the acceleration of gravity, C_d the drag coefficient, A the frontal area, ρ the air density and v the speed. Note that F_r is independent of speed while F_d is proportional to the speed squared (cf. Bernoulli's law). For an average car the rolling resistance dominates at speeds up to about 60 km/h, and the drag becomes dominant at higher speed.

The *rolling resistance* for tires is caused by the fact that the forces related to compression and expansion of rubber are not equal: there is some hysteresis. Note that the value of C_r must be on the order of 0.01, since a slope of approximately 1% suffices to get a car rolling. Extensive measurements [1] have shown that values of C_r vary from 0.007 to 0.014. Note also that F_r is proportional to the mass of the vehicle. Therefore, in urban traffic the mass is a major factor in energy use, the more so since the energy lost in the braking/accelerating cycles is also proportional to the mass.

The *drag coefficient* C_d depends totally on the shape of the vehicle. Bernoulli's law tells us that it must be around unity for a flat plate whose surface is perpendicular to the direction of motion. A number of values are given in table I.

1·1. *Cars.* – At speeds around 100 km/h it is found that the total resistance of an average car is around 500 N. So the mechanical energy use of a car on the highway is around 500 kJ/km. With an efficiency of 25% this means that the primary energy use (gasoline, for example) is around 2 MJ/km. Given an energy of combustion of 35 MJ/litre, we see that the energy use for driving on the highway can be expected to be around 18 km/L, which corresponds nicely to the real-life value.

The *power* needed for driving on the highway is found by multiplying the total resistance (500 N) by the speed (around 30 m/s), which yields approximately 15 kW.

1·2. *Buses, trains, aircraft.* – If the energy use for cars, buses and trains are compared, the following picture emerges. It is found [2, 3] that, per seat, buses are comparable to cars in city traffic, but more efficient in long-distance travel by a factor of 2 to 3. Trains can beat cars by a factor of 10 per seat, provided they do not stop frequently. An exception is the TGV (or high-speed train) whose energy use per seat is comparable to a car.

Aircraft form a special case since they have no rolling resistance once they are airborne. They are found to have a fuel consumption per seat which is about twice as large as for cars. Here, the disadvantage of very high speed is offset largely by the fact that the air density at cruising altitude is only one fourth of the density at sea level, combined with a low value of C_d and a relatively small frontal area per seat. An interesting feature for air travel is that long-haul flights carry a surprising amount of fuel. The mass of the fuel at take-off can be roughly equal to the mass of the empty plane. Put differently, it amounts to a fuel mass of 250 to 300 kg per passenger.

2. – Future energy carriers and sources

Transportation requires a convenient mobile energy carrier. For most applications liquid fuels like gasoline, diesel or kerosene are exceptionally well suited, since their energy density is second to none. The question is which energy carriers will be most suited to replace oil-based fuels after the fossil-fuel era. The most obvious options seem to be biofuels, batteries (in combination with supercapacitors), hydrogen and metal fuels. But also solar power must be considered.

2·1. *Biofuels.* – This does not seem a viable option, for several reasons. First, the efficiency of photosynthesis is rather poor: around 0.5 to 1% in moderate-climate regions, like in Germany for example, up to about 3% in favorable climates. This means that "first generation biofuels", which are derived from food products like corn, sugar or palm oil, are in competition with the world food supply ("food for the poor *vs.* fuel for the rich"). Second, if the losses in conversion and changes in land use are also considered, it is found that the CO_2 emission of biofuel-powered transport is at present comparable to that of fossil fuels, or even larger for "first generation biofuels" [4, 5].

2·2. *Batteries and supercapacitors.* – Electric vehicles (EVs) have many advantages over vehicles powered by an internal-combustion engine. Electric motors are light and relatively simple, and they do not have to idle when the vehicle is not moving. Moreover they deliver maximum torque immediately, and it is easy to save energy by regenerative braking.

One should distinguish hybrid cars, which ultimately run on gasoline or diesel; plug-in hybrids, which run partially on electricity; and real electric vehicles (EVs) which run exclusively on electricity. We will consider EVs in this chapter.

There is little doubt that electric vehicles will be the leading kind of car in the distant future. In the present transition period, however, the advantages of electric cars are marginal. The CO_2 emission, for example, may not be lower than for a car powered

TABLE II. – *The energy density of various media for storing electricity. For comparison: the energy density of gasoline is about* 40 MJ/kg *or* 11 kWh/kg *in terms of thermal energy.*

Lead battery	≈ 0.03 kWh/kg
Ni-MH battery	≈ 0.08 kWh/kg
Li-ion polymer battery (LiPo)	≈ 0.20 kWh/kg
Li-ion battery	≈ 0.25 kWh/kg
Supercapacitor	≈ 0.005 kWh/kg
expected:	≈ 0.02 kWh/kg

by diesel or gasoline, depending on the way the electricity is generated. Let us take an EV using electricity generated in a coal-fired power plant. The CO_2 emission of coal combustion per unit of energy is 89 kg CO_2/GJ. If we take the net efficiency of the power plant in producing electricity to be 35% and consider some additional losses in the charge-discharge cycles of the battery plus a few percent in the electric motor, the final fuel-to-wheel efficiency can be estimated to be 30%. This means that the above CO_2 emission must be raised to 89/0.30 kg CO_2/GJ = 297 CO_2/GJ. The emission of a diesel-powered car having an efficiency of 25% compares favorably with this value: taking the specific emission of oil as 71 kg CO_2/GJ, we end up with an emission of 71/0.25 kg CO_2/GJ = 284 kg CO_2/GJ, which is even a bit less than it is for the electric car. An advantage of the EV is, of course, that the emission is at the power plant site rather than in urban areas.

Battery performance has been improving remarkably over the last years. The present (2016) energy density of Li-ion batteries is 0.25 kWh/kg [6]. This is much higher than, for example, traditional lead batteries, which have about 0.03 kWh/kg (see table II).

Even so, we should bear in mind that the energy density of batteries is still very low if compared to liquid fuels. One way to see this is to realize that a battery stack of 10 kg of Li-ion batteries contains the same amount of energy as a beer glass full of gasoline, the combustion energy of gasoline being about 36 kJ/L, or 10 kWh/L. The advantage of a battery, of course, is that the energy it contains can be converted into mechanical energy without any losses in principle, and with losses of only a few percent in practice. By contrast, the efficiency of an internal combustion engine powered by gasoline is only about 20–25%.

The price of Li-ion batteries at present (2016) is down to approximately 500 USD/kWh [6].

Whereas batteries are well suited for storing energy, their performance in producing high power is limited. Therefore, batteries are most often accompanied by super capacitors (or "ultracapacitors" or "double layer capacitors"), because these can produce high power, being well suited for frequent charge and discharge cycles at high current and short duration. This is obviously important for regenerative breaking and rapid acceleration. One should realize that such capacitors have capacitances on the order of farads, or even kilofarads. This is many orders of magnitude larger than the traditional capacitors

TABLE III. – *Two Electric Vehicle types compared: EV1, a relatively light car having a small maximum range, and EV2, a heavy car having a large maximum range.*

	EV1	EV2
Mass	1195 kg	2085 kg
Electric motor max. power	125 kW	568 kW
Battery (Li-ion)	19 kWh	85 kWh
Energy use per 100 km	14–17 kWh	24 kWh
Maximum range	130–160 km	426 km (EPA)

in electronic devices, which have capacitances on the order of picofarads or microfarads. Supercapacitors can store typically 5 Wh/kg, which is poor compared to 250 Wh/kg for Li-ion batteries, but they can deliver much higher powers: up to or exceeding 10 kW/kg. An interesting comparison between batteries and capacitors was given in Physics Today, December 2008 [7].

Given the limited energy density of batteries in terms of the number of GJ per kg, the maximum range of an EV is still an issue. In this context one may choose between two concepts: either opt for a relatively light EV having a small range and suited for example for commuter traffic, or a relatively heavy car having a large range. In table III two such cars are compared: EV1 (a BMW i3) and EV2 (a Tesla model S).

It is seen from table III that a large range comes at a price: EV2 has almost double the mass compared to EV1, and accordingly its energy use is higher. This illustrates that contemporary EVs should be chosen according to the prospective use regarding desired range and charge opportunities.

A drawback of EVs at present is the long charging time. This is easy to understand if we realize that the power for driving is around 15 kW (see sect. **1**˙1). By comparison, the maximum power for a standard electrical outlet in a European home is limited to about 3.5 kW given a voltage of 230 V and a 16 A fuse. This means that for every hour driving on the highway we need a charging time of four hours, if we assume charging from a standard home outlet. It is illustrative to compare this with refueling gasoline. This is done at a rate of about 0.6 Liter per second. This corresponds to about 20 MJ/s or 20 MW! Another drawback of EVs in winter time is that heating the car is electric, so it goes at the expense of battery charge. For a car having an internal combustion engine heating is free, since it is simply extracted from the waste heat.

2˙3. *Hydrogen.* – As a future energy carrier hydrogen seems to be rather ideal. It can be manufactured for example by using electricity via electrolysis with an efficiency of up to about 80%. It can also be produced directly from sunlight by some green algae and cyanobacteria [8]. Although the production of this so-called photobiological hydrogen is still in an experimental phase with efficiencies of about 2%, this development may become important in the future. Once produced, hydrogen can be used to generate electricity in a fuel cell with an efficiency of up to about 60% so far. The problem is to store hydrogen

TABLE IV. – *Properties of molecular hydrogen* (H_2).

Heat of combustion (upper[(a)])	$12.8\,MJ/m^3$ or $142\,MJ/kg$
Heat of combustion (lower[(a)])	$10.8\,MJ/m^3$ or $120\,MJ/kg$
Density (at 0 °C, 1 bar)	$0.090\,kg/m^3$
Boiling point	$-253\,°C$ or $20.4\,K$
Density of liquid H_2	$71.0\,kg/m^3$
Latent heat of vaporization	$449\,kJ/kg$ or $31.8\,MJ/m^3{}_{liquid}$

[(a)] The "upper" value includes the heat of condensation of the water vapour; the "lower" value does not.

in a form suitable for mobile use, since it is in a gaseous state at ambient temperatures. One should bear in mind that the density of gases at normal temperature and pressure is three orders of magnitude lower than for liquids.

The contemporary way to circumvent this storage problem is to compress the gas to pressures of, for example, 700 bar. In addition to requiring strong yet light and small containers, this has its limitations which are inherent to the hydrogen molecule. Since it is a light molecule having only two protons and two electrons, the attractive forces between molecules are small. The first consequence is that it has a very low boiling point: 20.4 K (see table IV).

The second consequence is that the repulsive forces are by far dominant at ambient temperature. This means that when compressing the gas, the pressure increases more than one would expect on the basis of the ideal-gas law. For example: at 700 bar the density is only 2/3 of what one would naively expect. The third consequence is that, in the liquid phase, the attractive well is so shallow that it is relatively easy for the molecules to escape into the gaseous phase. In other words: the heat of vaporization is relatively small: it is only 31.8 kJ/L (or 0.90 kJ/mol), compared to 2400 kJ/L (or 43 kJ/mol) for water. This low heat of vaporization is a drawback when storing hydrogen in the liquid phase: one needs very good thermal insulation if the boil-off is to be small.

This inevitable boil-off, combined with the fact that liquid hydrogen is more difficult to handle than ordinary fuels like gasoline, makes the liquid option not very suitable for cars. However, there may be a great future for the use of liquid hydrogen in aviation. There are several reasons for this. The first reason is that weight is an important issue in aviation. This is especially the case for long-haul flights, where the weight of the kerosene is comparable to the weight of the empty aircraft (see sect. **1**·2). The weight of hydrogen per unit of energy is about a factor of 3 lower than for kerosene. The second reason is that the boil-off is much less of a problem, not only because the aircraft is airborne only for a limited amount of time, but also because the temperature difference between the liquid hydrogen and the outside air is much smaller at cruising altitude than it is at ambient temperature. The third reason is that refueling of aircraft is done by professionals only, which reduces the risk of accidents to a level which is comparable to the risk involved in handling kerosene.

For the sake of completeness we also mention the option of storing hydrogen as a metal hydride. Some metals are capable of absorbing large amounts of hydrogen in atomic form, even to the extent that the amount of hydrogen per unit volume is larger than in the liquid phase. However, the disadvantage is that the "tank" is very heavy compared to the weight of the hydrogen inside. This is obvious because hydrogen is the lightest of the elements and metals tend to be heavy. An example of a metal-hydrogen compound ("metal hydride") is $LaNi_5H_6$, where hydrogen is stored in an alloy of lanthanum and nickel, both heavy metals. If we look up the masses of La and Ni, we find that the mass ratio of hydrogen to metal is 6 to 432. But the good news is that 6 kg of hydrogen enables us to travel 400 to 800 km, depending on whether a combustion engine or a fuel cell is being used.

2˙4. *Metal fuels.* – A recent development is the option of metal fuels [9]. This option can be completely CO_2-emission free. The idea is to equip a car with a cartridge containing, for example, a few liters of porous micron-sized iron powder. When driving the car, the iron powder is dispersed in air and the powder is burned in a hot turbulent flame, converting the iron powder into micron-sized rust-powder. The combustion energy is then used to drive a Stirling engine. The burnt rust particles are stored in a separate compartment of the original cartridge and when returning home the (heavier) cartridge is brought back to the store, from where it is shipped to a central place where the rust is transferred back to micron-sized iron particles using some clean primary energy, like solar energy. A clear advantage is that the energy density of iron is rather high: about 60 MJ/L. Note that metal fuels can be directly burned with air.

2˙5. *Solar power for transportation.* – Designing an experimental racing vehicle powered by solar energy is relatively easy, as is demonstrated in the biannual World Solar Challenge race across Australia. An example of such a car is shown in fig. 1.

By contrast, it is quite a challenge to design a family car powered by solar energy. A group of students at the Eindhoven University of Technology in The Netherlands took up this challenge, and won the Cruiser Class of the World Solar Challenge 2013. They built a family car which can seat 4 persons and has a range of up to 1000 km. The specifications of an updated version are shown in table V.

This car, the Stella Lux, can be driven without the use of batteries under a clear sky at a speed of about 43 km/h. In practice also use Li-ion batteries having a capacity of 15 kWh are used, which enable the car to have a range of 1000 km on a sunny day in The Netherlands, and about 650 km at night. The top speed is 125 km/h. But this comes at a price: the comfort is bound to be marginal (see fig. 2).

Can this concept be used also for aviation? After all, in the summer of 2016 a solar-powered airplane, the Solar Impulse 2, flew from New York City to Sevilla in Spain. It had 270 m^2 of solar panels, delivering 66 kW peak. It used Li-ion batteries having a total energy of 164 kWh and having a mass of 633 kg. Granted, the speed was not particularly impressive, being 90 km/h during day and 60 km/h at night (to save energy), but it is only a start. Could this be scaled up to a commercial airliner comparable to, say,

Fig. 1. – Solar racing car *Nuna*, winner of the World Solar Challenge in Australia. Using a $6\,\text{m}^2$ area of solar cells having an efficiency over 26% and a 1.5 kW electric motor, its top speed was over 100 km/h. The mass was 190 kg and its rolling resistance only about 1/10 of that of a normal car.

a Boeing 747, with a reasonable speed? Alas, this is not the case, and it is easy to see why. The Boeing 747 has an engine power of about 110 MW (which, by the way, is on the order of a small power plant!). Its wing area is $550\,\text{m}^2$. Let us suppose we cover this area with solar panels having an efficiency of 20%. We recall that the solar irradiance is almost exactly $1\,\text{kW/m}^2$, so that the maximum electrical power delivered by the total wing surface is 110 kW. This happens to be three orders of magnitude smaller than the actual power mentioned above. So, unless revolutionary changes will take place in the plane concept, the idea of solar-powered passenger planes seems beyond hope.

TABLE V. – *Technical specifications of the solar-powered family car "Stella Lux".*

Number of seats	4
Height	1.12 m
Mass	375 kg
Battery (Li-ion)	15 kWh
Solar array (monocrystalline Si, 23% efficiency)	$5.8\,\text{m}^2$
Power at 100 km/h	≈ 1.5 kW
Range on sunny day (Netherlands)	1000 km
Range by night	≈ 650 km
Top speed/cruising speed	125/70 km/h

Fig. 2. – The solar-powered family car Stella Lux, built by a group of students at Eindhoven University of Technology, Eindhoven, The Netherlands (see table V). Courtesy: Solar Team Eindhoven.

3. – The bicycle

Of course the most sustainable means of transportation are those which are powered by human energy. Walking is relatively inefficient if it is compared to cycling. This

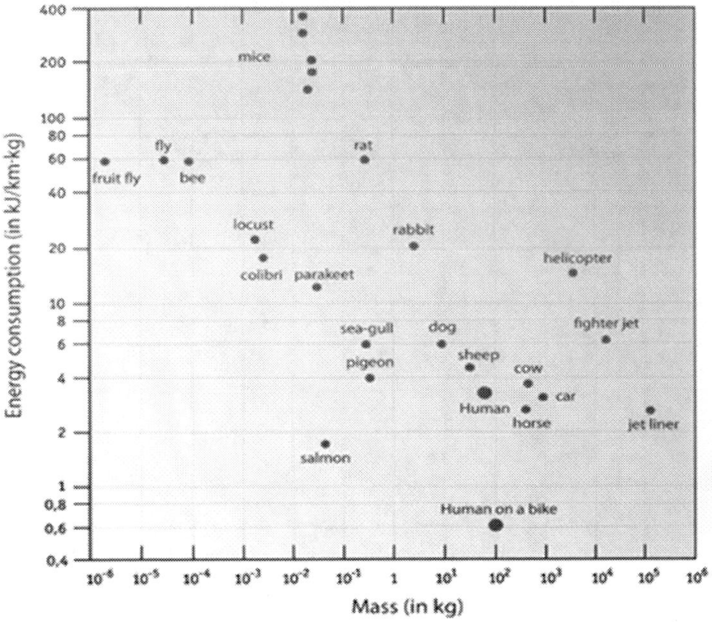

Fig. 3. – Energy use per km and per kg for various modes of transportation, arranged from low mass (left) to high mass (right).

can be readily understood, since cycling increases both the speed and the distance to be covered in a day by a factor of 4 to 5 with the same effort. This is also nicely illustrated in fig. 3 where various modes of transportation are compared. It is seen that a human being on a bicycle is the most efficient in terms of energy use per km and per kg.

The number shown for a bicycle is slightly arbitrary, since it depends heavily on speed and on the sort of bicycle. To see this we turn to fig. 4, where the resistance for a cyclist is shown as a function of speed. It is seen that, for an ordinary city bike, the total resistance (and, consequently, the energy use per km) goes up dramatically at higher speeds. This is because the air resistance (or drag) dominates above about 15 km/h. Consequently, the energy use per km also goes up dramatically at high speeds. The only way to reduce this is to decrease the aerodynamic drag, for example by assuming a streamlined, crouched position rather than sitting upright. This is pushed to the limit by Human Powered Vehicles (HPVs), in which the rider is often lying horizontally, and which have values of the drag coefficient C_d of around 0.1 instead of 1, which is the value for a cyclist sitting upright on a city bike shown in fig. 4. This enables HPV riders to reach speeds far above 100 km/h, as demonstrated, for example, in 2015 by Todd Reichert, who rode his vehicle at a record speed of 139.45 km/h. Such high speeds are most often reached at elevated altitude (on Battle Mountain, Nevada in this case) to reduce drag by lowering the air density.

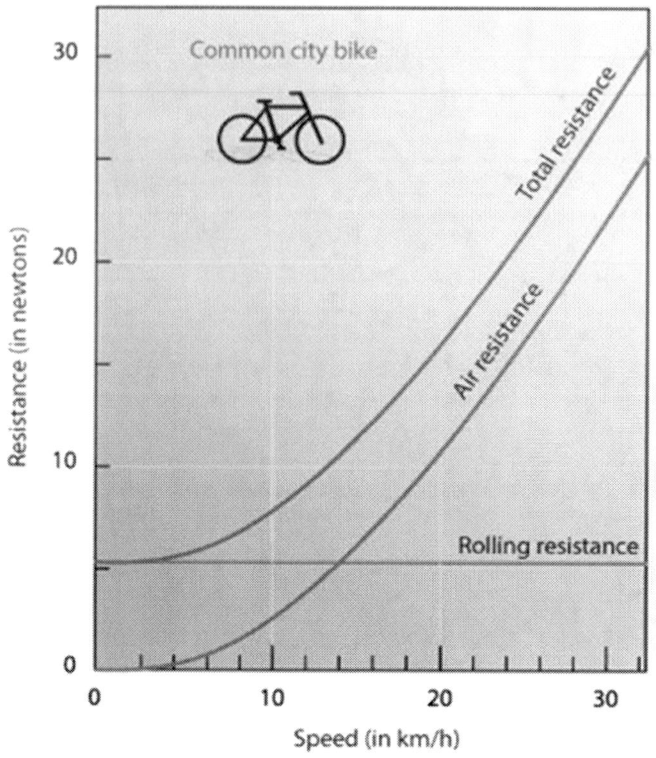

Fig. 4. – Rolling resistance and air resistance (or aerodynamic drag) for a city bike and a rider sitting upright. It is seen that drag is dominant above about 15 km/h.

4. – Conclusion

Achieving sustainable transport systems in the post-fossil-fuel era will pose a real challenge, because of the crucial role of liquid fuels in this sector. For cars and buses, batteries in combination with supercapacitors can provide a good alternative. Hydrogen in compressed form may also be an option, especially if used in combination with fuel cells. For aviation, some sort of liquid fuel with a high energy density seems to be indispensable in view of weight limitations. In this case liquid hydrogen may provide an excellent option for a number of reasons.

Direct use of solar power may be viable in some cases. Solar-powered vehicles are within reach, provided that customers are willing to accept a lower degree of comfort and speed. Solar-powered aviation, by contrast, seems beyond hope. And in view of the high efficiency of riding a bicycle, an excellent motto may be: cycle … and recycle.

REFERENCES

[1] National Academy of Sciences, *Tires and Passenger Vehicle Fuel Economy* (2006). Available at http://onlinepubs.trb.org/onlinepubs/sr/sr286.pdf; https://www.nap.edu/catalog/11620/tires-and-passenger-vehicle-fuel-economy-informing-consumers-improving-performance.

[2] Hermans J., *Energy Survival Guide* (Leiden University Press, Leiden, The Netherlands) 2011.

[3] Cifarelli L. and Wagner F. (Editors), *Energy: Basic Concepts and Forefront Ideas*, in *Lecture Notes Joint EPS-SIF International School on Energy* (SIF, Bologna) 2015, p. 339.

[4] *Lifecycle analysis by Transport & Environment*, based on Globiom Study (2016), accessible by https://www.transportenvironment.org/sites/te/files/publications/2016%2004%20Globiom%20webinar.pdf.

[5] Ecofys, IIASA and E4tech, *The land use change impact of biofuels consumed in the EU, Quantification of area and greenhouse gas impacts* (2015), accessible by https://ec.europa.eu/energy/sites/ener/files/documents/Final%20Report_GLOBIOM_publication.pdf.

[6] Crabtree George, talk presented at *the 69th Course of the International School of Solid State Physics: Materials for Energy and Sustainability V, Erice, Italy, 2016.*

[7] Abruna H. D., Kiya Y. and Henderson J. C., *Phys. Today*, **61**, issue No. 12 (2008) 43.

[8] De Groot H., private communication.

[9] Deen N., Bergthorson J., Shoshin Y. and Van Der Schaaf John, *Energy Days, 27 October 2016, Eindhoven Energy Institute* (Eindhoven University of Technology, The Netherlands) 2016.

Inorganic materials for photovoltaics: Status and futures challenges

A. SLAOUI(*)

Laboratoire ICUBE, CNRS & University of Strasbourg
23 rue du lœss, 67037 Strasbourg, France

Summary. — This paper review the present technologies for the fabrication of solar cells and modules based on the most common semiconductors namely silicon, CuInGaSe(S) and CdTe materials as well as on III-V concentrated photovoltaic cells and modules. For silion technology, we give insights on the growth of monocrystalline and multicrystalline silicon wafers and then we describe the most common solar cells designs and how to fabricate them. We also provide information about the fabrication of silicon modules and their performances. As for the thin-films solar cells, we present the structurale and optical properties of the CIGS and CdTe materials as well as the solar cell structures. The multi-junction concept cell that involves several III-V materials of different bandgaps is also described, and data on their fabrication, performances and mounting as modules are presented. Finally, a short outlook on the coming materials for solar cells is provided.

1. – Introduction

The technologies of photovoltaic cells and modules are numerous and call for diverse and varied materials and manufacturing processes [1, 2]. They are traditionally classi-

(*) E-mail: `Abdelilah.slaoui@unistra.fr`

fied into three categories related to their maturity: first-, second- and third-generation modules, the distinctions between these generations depends on the author! (fig. 1).

The so-called first-generation technologies include cells and modules made from wafers. This is the case for crystalline silicon and tandems made from III-V elements. Crystalline silicon (sc-Si) technologies have dominated the market since the early 1960s [3, 4]. Automation, upscaling, better light management, and Chinese manufacturing capabilities have lowered the cost of these modules to a level that it is making it difficult nowadays for competing technologies to gain market shares. This first generation of solar cells is now mature and the main improvement paths consist, on the one hand, of increasing the efficiency by using n-type silicon and by carrying out heterostructures with amorphous silicon or even perovskites and on the other hand, to reduce manufacturing costs by constructing "Giga-Factories" and reducing the silicon cell thickness. Besides, multi-junction cells based on III-V compounds have traditionally been developed for space applications since the 1990s and have been used for several years in concentration systems.

The second generation consists of thin-film modules of simple semiconductor materials (amorphous and micro or nano-crystalline silicon) [5] or compounds (CIGS and CdTe), whose manufacturing processes involve depositing a thin layer of active and absorbing semiconductor on an inexpensive substrate (glass, metal or polymer). Modules with amorphous silicon (a-Si) have been manufactured since the 1980s, whereas the CdTe ad CIGS modules appeared at the mid-2000s. For long time, the industry considered the thin films as a real technological breakthrough capable of taking a significant market share. However, the growth of the CdTe and CIGS sectors is delayed by the very low price of the crystalline silicon modules. The amorphous silicon based modules are currently experiencing problems of survival because its efficiencies are low and its cost remains high. It will not be described in detail here.

The third generation of solar cells brings together technologies that are under development that can mature in the medium to long term by lowering costs and/or increasing efficiencies [6]. It includes i) advanced architectures to achieve efficiencies that can theoretically exceed 60%, ii) organic or hybrid thin films and iii) perovskites. For instance, thin hybrid films (a combination of inorganic and organic) were explored since the 90s, and the most studied structure was the so called "Grätzel" dye solar cells (or DSSC) that combine nanocrystalline anastase TiO_2 with organic dyes [7-10]. Conversion efficiencies, initially very low, have taken off these last six years due to the use of perovskites as absorber materials. However, these compounds have a limited lifetime (several thousand hours) linked to the degradability of organic compounds by atmospheric agents (UV, O2, water vapor) and to the instability of the interfaces between the organic active layer and inorganic electrodes such as conductive transparent oxides or metals. One of the challenges will not only be to stabilize performance but also to find encapsulation methods to achieve appropriate lifetimes for the production of mass electricity. We will not report on those structures below.

In the following, we will describe the technologies for the fabrication of silicon modules, CIGS and CdTe modules as well as III-V concentrated photovoltaic cells.

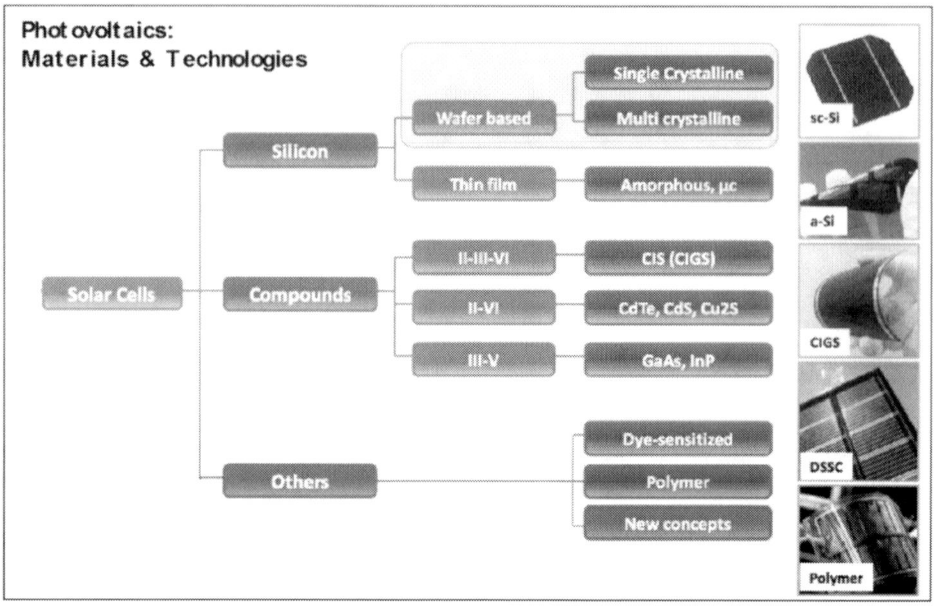

Fig. 1. – Materials and technologies used for photovoltaics.

2. – Crystalline silicon modules

The silica (SiO_2) found in the form of quartz is the basic material for producing silicon wafers for electronic and photovoltaic devices. The silica is reduced by carbon in an arc furnace to produce metallurgical silicon which, in turn, will have to be purified to obtain very high-purity silicon, called polysilicon. The latter is melted and solidified to produce ingots which can have different crystalline qualities according to the process used: monocrystalline silicon is obtained by drawing Czochralski or multicrystalline silicon by directional solidification. The ingots are then cut into bricks which themselves are sawn into thin wafers. The wafers make it possible to manufacture cells which are interconnected in series in sealed and mechanically resistant modules. The value chain for monocrystalline and multicrystalline silicon modules is shown in fig. 2.

2·1. *Single crystal silicon lingots*. – The monocrystalline silicon ingots are produced in most cases by the Czochralski (Cz) method. The polysilicon rocks are melted in a quartz crucible where they are placed in the presence of boron for the p doping of silicon with a typical concentration of 5×10^{15} to $10^{16}\,\mathrm{cm}^{-3}$ to reach resistivities of the order of 0.1 to $1\,\Omega\cdot\mathrm{cm}$. A monocrystalline seed is then quenched in the bath of liquid silicon and removed very slowly by a simultaneous movement of rotation and translation, resulting in the growth of an ingot from the bath. For a material of very good crystalline quality, it is necessary to control the growth temperature ($\approx 1410\,^{\circ}\mathrm{C}$), its uniformity and the speed of growth which should be slow and constant. The ingots are cylindrical due to the rota-

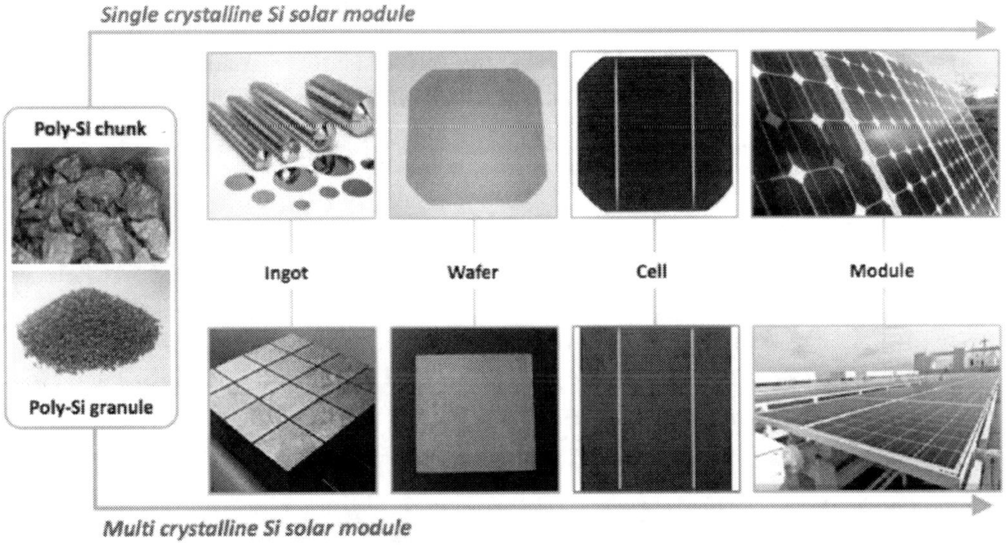

Fig. 2. – The value chain for the fabrication of monocrystalline (top) and multicrystalline (bottom) silicon photovoltaic modules.

tional movement. They have a diameter of between 10 and 30 cm nowadays, a length of 1 to 2 meters and a mass of 50 to 100 kg. The crystallographic orientation is controlled by that of the silicon seed and it us usually following planes {100} perpendicular to the direction of growth because this orientation makes it easy to texture the surface of the cells.

An alternative to the Czochralski method is the zone melting recrystallization technique (Fz). In this case, a very high-purity silicon bar is melted locally by induction, while a single crystal of silicon is extracted from this molten zone which moves in the opposite direction. These materials are exceptionally pure, in particular with regard to the very low oxygen content, insofar as the liquid silicon is not in contact with any wall and therefore there is no contamination by the crucible. Fz-Si is generally used in R & D where it has led to record efficiencies of crystalline silicon-based cells. The disadvantages of this technique for possible industrialization are, on the one hand, a higher production costs than those of Cz-Si and, on the other hand, the need to start from high-purity silicon bars, in contrast to Cz-Si which can be made from melted rocks.

2˙2. *Multicristalline silicon lingots.* – In order to reduce manufacturing costs, major research efforts began in the 1970s on the production of multi-crystal ingots (see bottom of fig. 2). The pieces of polysilicon are poured and then melted in a crucible made of silicon nitride or silica, of parallelepiped shape coated with SiO/SiN. Controlled cooling of the bottom of the crucible causes directional crystallization (from bottom to top) and the formation of a silicon ingot composed of large grains whose size is a function of the speed and direction of crystallization. The micro-structure is generally columnar, with grains up to a length of 10 to 20 cm and a section of the order of cm^2. This progressive

DOI: 10.1051/epjconf/201714800007

solidification, which maintains in contact with a solid phase and a liquid phase, develops a segregation of the impurities contained in the charge, which are drained upwards from the molded ingot. Parallelepipedal ingots weighing several hundred kilograms (currently 300 to 600 kg) are commonly prepared and are then cut into "bricks", which themselves are cut into plates.

The production of multi-crystalline silicon ingots is cheaper than the Czochralski ingots with less energy consumption. It does not require elaborate equipment, nor precise control of temperatures and speeds. Furthermore, since the process tolerates the basic impurities better, a less expensive silicon source can be used. The drawback is that the resulting material is of lower quality. The lifetimes of the carriers are lower and the intrinsic performances of the multicrystalline silicon-based cells are approximately 20% lower than those of monocrystalline silicon. This is due, among other things, to grain boundaries where impurities are accumulated, in addition to some crystal defects (dislocations) and the presence of metallic impurities.

2·3. *Fabrication of silicon solar cells.* – The basic material for the manufacture of the cells is a monocrystaline or multicrystamline silicon wafer, usually p-type, with a thickness of about 150–300 μm and a dimension of $156 \times 156\,\mathrm{mm}^2$. The conventional manufacturing process can be broken down into several steps (fig. 3). It can be noticed that the manufacture of cells based on crystalline silicon uses to a large extent technologies originally developed for the electronics industry: screen printing machines, drying and annealing furnaces for the application of contacts. The main steps are the following:

i) Surface treatment and texturing: an acid attack on the surface of the platelets is made in order to eliminate the damage due to sawing and to texture the surface and thus improve the absorption of the incident light.

ii) Formation of the PN junction by thermal diffusion of the dopants: the emitter is obtained by a diffusion of phosphorus limited to 0.2–0.3 μm of depth to create a n^+ highly doped emitter.

iii) Opening of the junction by reactive ion etching ($CF_4 + O_2$).

iv) Deposition of an antireflective layer of hydrogenated silicon nitride (SiN_x: H) on the front face. This deposition reduces the reflection coefficient in the center of the visible spectrum from 34% to about 8% and also serves as a hydrogen reservoir for passivating the emitter surface and the volume defects of the wafer.

v) Formation of the contacts by screen printing followed by a high-temperature annealing ($> 800\,^\circ$C): silver grid on the front and aluminum layer on the back.

2·4. *Performances of silicon solar cells.* – The efficiency of the cells obtained in industrial production according to the process described above is in the range of 18% for a monocrystalline substrate and 16.5% for a multicrystalline substrate. In the laboratory, the efficiencies obtained are much higher thanks to the use of microelectronics technologies.

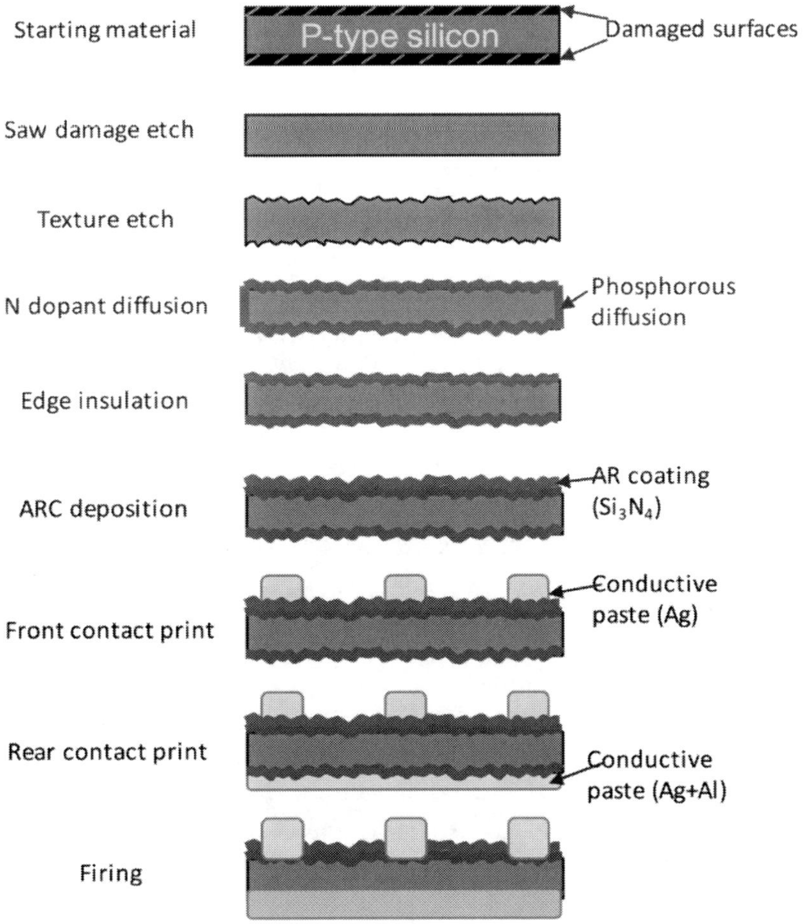

Fig. 3. – The different steps for the fabrication of a conventional silicon solar cell.

The world record for a silicon cell on a *p*-type substrate is held by the Center for Photovoltaic Materials and Systems at the University of New South Wales (UNSW) in Australia with a 24.7% Structure PERL (Passivated Emitter, Rear Locally diffused cell).

This structure made on a *p*-type FZ substrate (fig. 4) has several peculiarities compared to the standard process. The front is textured in reverse pyramids. The emitter is selective. A thin layer of thermal oxide is formed on the emitter to passivate the surface, and a double antireflection layer is deposited to reduce reflectivity. The back side is passivated by a thermal oxide which is then opened locally by chemical etching to make the contact. The contact region is previously doped p^+ in order to reduce the contact resistances and to realize the BSF. It should also be noted that the doping on the front and rear surfaces is done by ion implantation, this technique allows a perfect control of the dopant profiles.

Considerable progress has been made in the manufacturing process over the last fifty years. The thickness of the wafers, which was 400 μm in 1990, has gradually decreased to

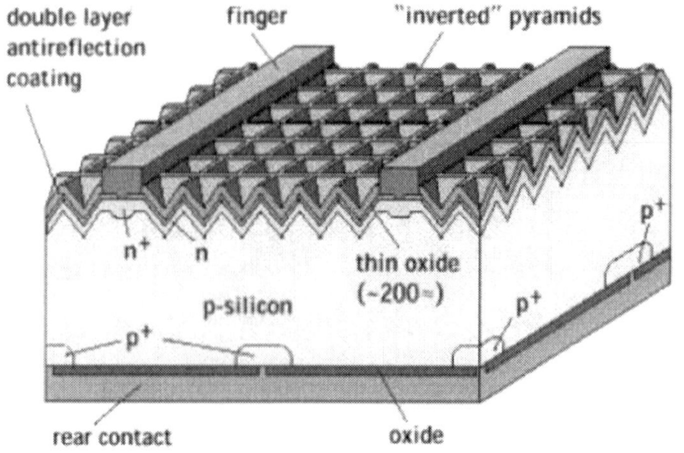

Fig. 4. – PERL structure as fabricated by University of New South Wales (Australia).

150–200 μm today. At the same time, the cell surface increased from 100 cm^2 to 240 cm^2. The output of screen-printed modules has increased from about 10% in 1990 to about 18% today for a monocrystalline substrate and 16.5% for a multi-crystal substrate, with the best modules exceeding 20%. The annual production of manufacturing plants increases from 1–5 MWp in 1990 to several GWp. The learning curve of wafer-based silicon technology shows over the last thirty years that the price of modules has dropped by 20% on average for each doubling of installed capacity. This decline is due to the combined effect of technological innovation, market development and the emergence of Chinese industry. This cost can be further reduced because each step of the manufacturing process can be further improved by strengthening the automation of manufacturing units, in particular by pursuing the paths taken in previous years to reduce the costs of the industry:

- Reduction of the amount of silicon used.
- Use of n-type monocrystalline silicon at back junction or in heterojunction with amorphous silicon which achieves high efficiencies.
- Reduction or even suppression of the metallization of the front panel by means of rear junction cells and/or rear contacts.

2·4.1. Rear junction cells. Conventional solar cells have contacts on both sides. In the rear-junction cell based usually on n-type monocrystalline cells (also known as interdigitated contact cells), the two contact are placed on the rear face and form an interdigitated grid allowing the metallization of the front face and the concomitant shading to be completely eliminated. The junction is also on the rear side and the emitter and the p^+ zones of the rear back surface field also form an interdigitated zone.

In conventional silicon cells, photogeneration occurs mainly on the surface. Here, the photogenerated carriers must therefore travel a distance which may be much greater than the thickness of the substrate to reach the junction at the rear of the cell. They can

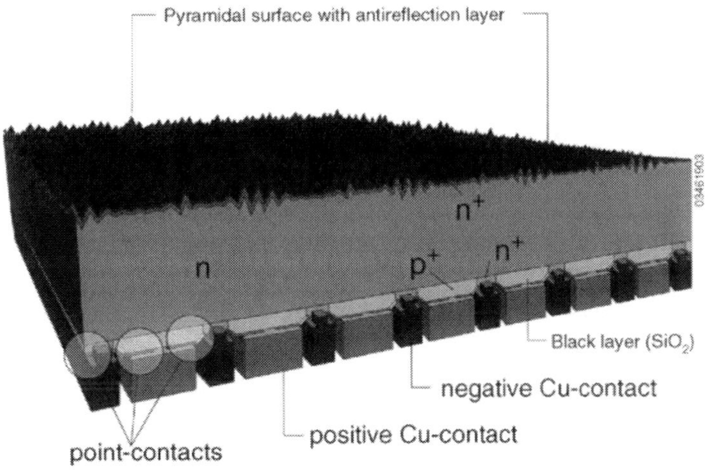

Fig. 5. – A "Sunpower" Interdigitated back contact cell based on type n-type silicon (source Granek).

recombine on the front if it is not well passivated or in the volume if their lifetime is not enough. The two most important parameters are thus the ratio of the diffusion length of the minority carriers to the thickness of the substrate, and the rate of recombination on the front face. Making a cell of this type therefore requires the use of high-purity monocrystalline silicon and high-quality manufacturing processes. They are reserved for uses where high efficiencies are suited. The rear junction cell A-300 marketed by Sunpower (fig. 5) was originally designed at Stanford University. It was marketed from 2004. In 2007, a new generation of A-300 cells reached an average yield of 22.4% in production. These cells have the highest production record. In 2010, a record 24.2% was achieved in R & D.

2·4.2. Heterojunctions: bifacial HIT cell. The HIT (Heterojunction with Intrinsic Thin Layer) cell on an n-type monocrystalline silicon substrate was designed by Sanyo Electronic Corporation and marketed since 1997. Its principle is based on the deposition of a two-layer stack of hydrogenated amorphous silicon (An intrinsic layer and a doped layer) on each of the faces of the wafer by chemical vapor-phase (PECVD) at temperatures ranging from 200 °C to 400 °C. The amorphous silicon forms a double tandem with the crystalline silicon (see fig. 6): On the front face, a p-i-n heterojunction was formed by successive deposition of an intrinsic hydrogenated amorphous silicon (a-Si: H) layer on the n-type silicon and then a p-doped amorphous silicon layer. The intrinsic layer provides excellent passivation of the crystalline silicon while the doped layer acts as an emitter. This cell design eliminates the step of forming the emitter by diffusion of boron; on the rear, the same structure is found. The last layer of amorphous silicon is highly doped n^+, and acts as a back surface field. Given the low conductivity of the amorphous silicon, it is necessary to add a transparent conductive layer (TCO) in addition to the

DOI: 10.1051/epjconf/201714800007

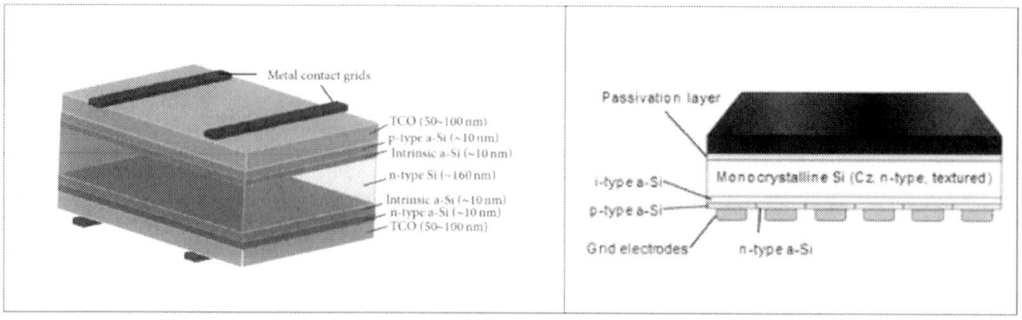

Fig. 6. – Schemes of a bifacial heterojunction cell (left) and back-contacts heterojunction cell (right).

silver collecting grids on each of the faces. This is illustrated in fig. 6. The structure obtained is symmetrical and can be manufactured as a bifacial cell which can allow an improvement of the efficiencies by taking part of the light scattered at the rear of the modules.

The next challenges are to reduce the thickness of the crystalline silicon wafer to 100 μm and to bring the contacts back to maximize the absorption of light. The efficiency world record on a heterojunction crystalline silicon cell (26.3%) has been held since 2016 by the Japanese company Kaneka Corporation on a small area (180 cm^2) cell with heterojunction on n-type substrate with back contacts.

2·5. *Silicon photovolatic module.* – The manufactured cells have a low open-circuit voltage (0.6 V) and are not useful as such. Modules are formed consisting of a number of solar cells of the same electrical characteristics which are interconnected (typically 72 in series) and encapsulated. A high voltage is thus obtained and the cells and their metallic contacts are protected from environmental stresses (moisture and hailstorms) and mechanical stresses (transport, assembly, wind) for a duration of at least 25 years.

The cells are generally connected in series in the module and are placed between 2 sheets of vinyl acetate (EVA), enclosed between two tempered glasses or between a glass plate and a tedlar sheet. A scheme of the module components is shown in fig. 7.

The glass used on the front of the module is a tempered glass with a low iron content. It must have a high transmission coefficient in the wavelengths used by solar cells (0.35 μm to 1.10 μm for silicon). The surface also needs little light reflection. In most cases, it ensures the mechanical strength of the module and the rigidity of the assembly. The glass also has the advantage of a perfect moisture tightness and a good self-cleaning power under the effect of rain. Its refractive index is close to 1.5, which is a good average between the air (index 1) and the cells (index 2 if they are provided with an antireflection layer). The two thin EVA sheets serve as a coating material and allow good adhesion between the cells and the front and rear faces of the module. The EVA has a good resistance to ultraviolet and temperature, a refractive index identical to that of glass and its implementation is relatively easy. The sandwich is laminated at 150 °C, with the

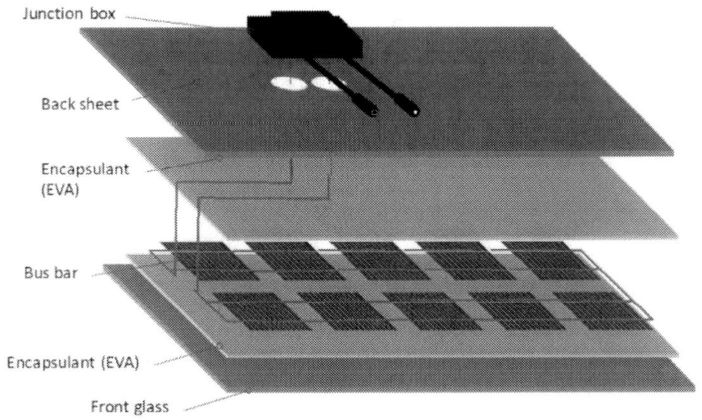

Fig. 7. – Scheme of a silicon module and its components.

aim of removing all the air from the module and crosslinking the EVA which will fill all interstices in order to give the material better resistance to temperature and UV.

The main function of the module's back is protection against moisture. Here a laminated sheet tedlar/aluminum/tedlar is generally preferred because lighter. Aluminum is a good barrier against moisture. One of the tedlar films isolates aluminum from the cells while the other protects aluminum from atmospheric conditions. The module is generally provided with an anodized aluminum frame which reinforces the strength and facilitates the fixing of the module on a structure. The back of the module is equipped with a connection box that contains the hot spot protection diodes and allows the module to be connected to the external circuit.

3. – Polycrystalline thin-film modules (CIGS and CdTe)

The family of thin-film compounds ($< 5\,\mu$m) of crystalline semiconductors consists of direct band-gap materials that are highly visible and highly tolerant to defects [11, 12]. The two most studied families are the CIGS chalcopyrites and the cadmium telluride (CdTe) [13-16]. The photovoltaic cells made with these materials are of the heterojunction type, that is to say that the p and n regions are made of different materials. The *p*-type semiconductor, $Cu(In, Ga)(Se, S)_2$ (or CIGS) or CdTe, is the absorber of the cell (*i.e.* the area where the electron-hole pairs are photogenerated). The junction is formed with a cadmium sulphide (CdS) emitter, a broad band gap doped semiconductor. The absorption is minimized in the transmitter by the choice of a high band gap (E_g (CdS) $= 2.42\,$eV). This layer passes the visible radiation which is then absorbed into the absorber. The carriers are therefore only photogenerated in the absorber unlike the silicon where the n layer contributes to the photocurrent.

All these thin-film technologies have the advantage of being able to be deposited on rigid supports such as glass, or flexible materials such as stainless steel or plastic sheets (*e.g.*, polyimides). The manufacturing processes used make it possible to avoid sawing of wafers and the inherent losses of raw material. Modules can be made into lightweight sheets or rolls and thus be more easily incorporated into building materials such as roofing elements, windows or facades. The conversion efficiencies obtained are still lower than those achieved with crystalline silicon, although significant progress has been made in recent years.

3·1. *Chalcopyrities (CIGS).* – The technology called CIGS corresponds to the manufacture of thin-film modules using chalcopyrite-type ternary compounds as an absorbent layer: $CuInSe_2$ (CIS), $CuInS_2$, $CuGaSe_2$ (CGS), with respectively bandgaps of 1.0 eV, 1.5 eV and 1.7 eV, and their multi-alloys $Cu(In, Ga)(Se, S)_2$ whose band gap can be made tunable depending on the composition. All these compounds have a direct band gap, which makes them suitable for use as an absorber in a thin-film photovoltaic cell because of their very high absorption coefficient (of the order of $10^5\,cm^{-1}$) in the visible range. This is compatible with the use of a thickness of 1 to 2 microns, about 100 times smaller than in the case of crystalline silicon. With this technology a record 22.6% efficiency has been achieved in the laboratory, while the typical yield of the marketed modules is in the order of 12 to 14%.

The compounds $Cu(In,Ga)(Se,S)_2$ belong to the so-called I-III-VI2 family. They crystallize in a structure derived from the Zinc-Blende structure. In the case of CIGS, this structure has an anionic sub-network (ions C_{2-}, *e.g.* Se_{2-}) and a cationic sub-network (ions A^+ and B^{3+}, *e.g.* Cu^+ and In^{3+}). The $CuInSe_2$ structure, for example, can be deduced from the cubic zinc-blende structure of an II-VI material such as ZnSe in which the sites of the Zn^{2+} cations are alternately occupied by the Cu^+ and In^{3+} cations.

The CIGS can have an electrical conductivity of type n or of type p, due to the presence of numerous intrinsic defects in the structure of the material. It is therefore possible to dope the CIGS with native defects without introducing extrinsic impurities as would be done with silicon. For example, an excess of selenium will lead to the p-type, a deficiency of selenium will lead to type n. The current photovoltaic cells use $Cu(In,Ga)(Se,S)_2$ films prepared with an excess of Se and are therefore p-type with predominant hole concentrations between 10^{16} and $10^{17}\,cm^{-3}$ and diffusion lengths of the minority electrons of The order of the micron. In addition, CIGS-based cells are stable over time due to their resistance to radiation due mainly to the mobility of copper and the high density of defects, which are rearranged and ensure electrical stability.

The properties of CIGS (band gap and mesh parameters) can be varied by substituting some of the atoms of indium and/or selenium with atoms of the same valence (Ga or Al for In, S for Se). We then have band gap energies that vary between 1.04 eV for $CuInSe_2$ and 2.4 eV for $CuGaS_2$.

Figure 8 (left) represents the bandgap as a function of the lattice constant parameter for different chalcopyrites. The best performances have been obtained by substituting a portion of the indium atoms with gallium atoms, resulting in materials of the formula $CuIn_{1-x}Ga_xSe_2$: an optimum yield is observed for $x = 0.2$–0.3 with $E_g = 1.12$–1.20 eV (fig. 8 (left)). Another way is to substitute selenium atoms with sulfur atoms, leading to materials whose bandgap varies between 1.04 eV and 1.57 eV.

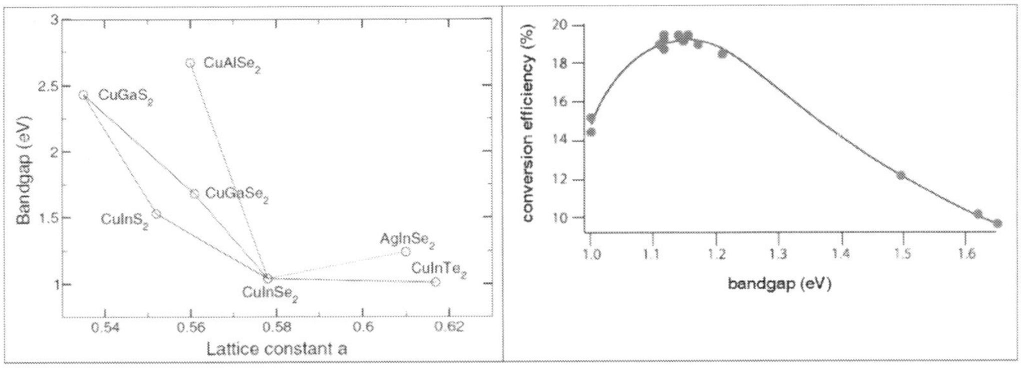

Fig. 8. – Left: bandgap *versus* lattice constant for some I-III-VI_2 chalcopyrites. Right: conversion efficiency of $CuIn_{1-x}Ga_xSe_2$, *versus* bandgap; the optimal efficiency is for $x = 0,3$.

The most used deposition techniques of the chalcopyrites active layer are co-evaporation or reactive cathodic co-sputtering at high temperatures of the Cu-In-Ga-Se-S elements which allow good control of the crystal growth of the CIGS material. They currently make it possible to obtain the best cell efficiencies. Another one is the cold deposition, followed by annealing and selenization and/or sulfurization. This is the case for sputtering, electrodeposit, ink jet printing techniques, which require annealing in order to obtain an absorber with sufficient crystallinity and good optoelectronic properties. On the other hand, processes at atmospheric pressure are attractive because they potentially greatly reduce initial investment compared to vacuum processes. However, the obtained module efficiencies remain lower.

The classical structure of a thin layer cell of $Cu(In,Ga)(S,Se)_2$ is described in fig. 9 (left). Figure 9 (right) shows a transverse section through a scanning electron microscope of a real device. The latter has five thin layers (fig. 9 (right)) deposited on a rigid soda-lime glass (1–3 mm thick), a metal or a flexible polymide substrate.

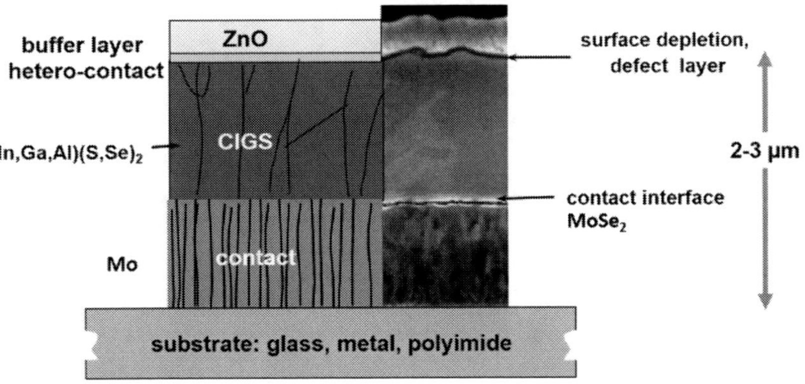

Fig. 9. – Left: scheme of a CIGS cell; right: MEB cross section of a CIGS structure (NREL).

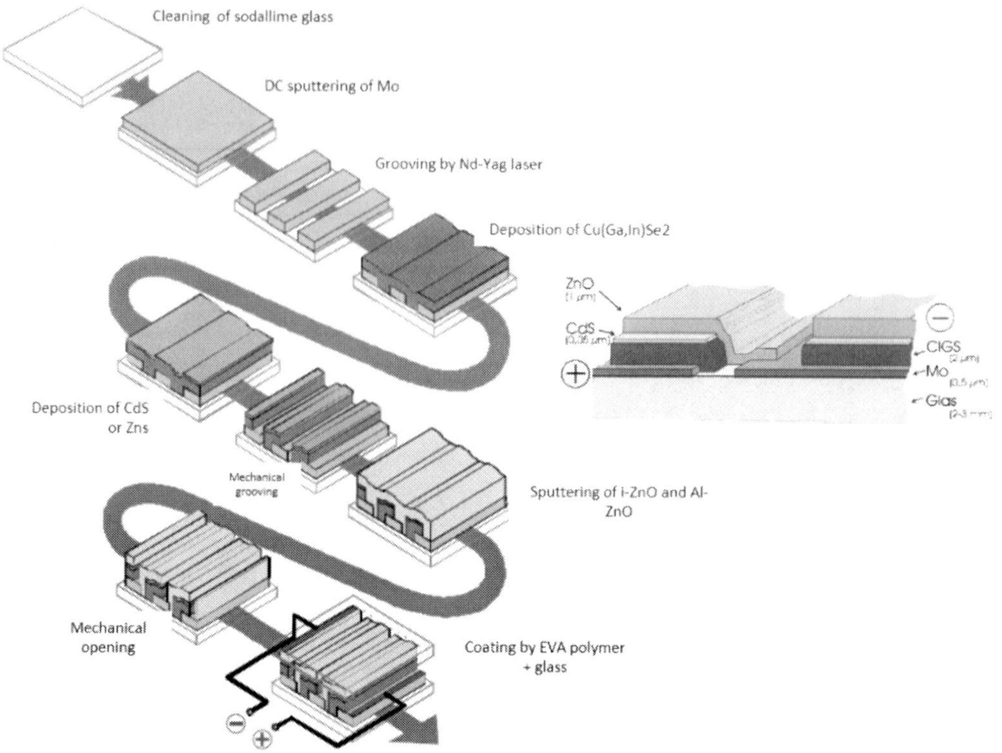

Fig. 10. – Left: steps for the fabrication of CIGS modules. Right: cross section of a CIGS cell.

The manufacture of CIGS modules differs from that of the conventional crystalline wafer silicon modules (fig. 10). The deposits are directly carried out on large surfaces (the sold modules have typically a surface area of $1.7\,m^2$). In order to avoid excessive resistive losses and to achieve a high open-circuit voltage, a series of etchings is carried out after the layers have been deposited, in order to define cells and to put them in series by a monolithic interconnection method.

3·2. *Cadmium telluride thin-film cells (CdTe).* – Cadmium telluride (CdTe) is a II-VI compound that has a bandgap of 1.44 eV and is therefore ideally suited for photovoltaic conversion. Its bandgap is of direct type and its absorption coefficient α (λ) for visible light is greater than $10^5\,cm^{-3}$. The absorbent layer therefore requires only 2 micrometers to absorb 99% of the solar spectrum without the need for sophisticated optical confinement.

Its absorption qualities and the wide variety of low-cost deposition techniques that can be used have made CdTe a particularly attractive material for the photovoltaic sector. CdTe's thin-film photovoltaic modules have been in development since the 1960s. They have been at the commercial stage since the mid-2000s. The American company First Solar is now the dominant player in this sector.

DOI: 10.1051/epjconf/201714800007

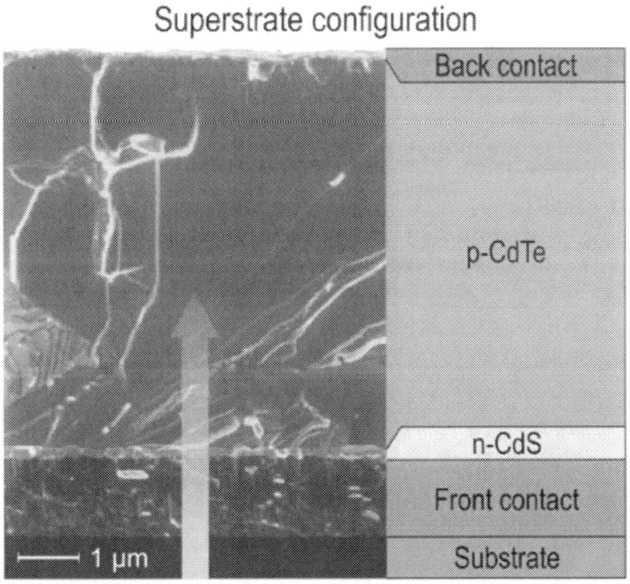

Fig. 11. – Left: MEB cross section of a CdTe based structure. Right: scheme of a CdTe cell in superstrate configuration.

A cross section of the CdTe cell structure is shown in fig. 11. The record laboratory yield was 22.1% while the average yield of the marketed modules was 16.1% at the end of 2015.

One of the advantages of CdTe thin-film modules is that they can make a monolithic interconnection of cells so as to achieve a high open-circuit voltage. The steps are similar to those described above for the CIGS modules.

4. – High-concentration photovoltaic systems

The maturity of multi-junction cells (also called tandem) based on III-V compounds, which were developed for space applications from the 1990s, opened up new opportunities for terrestrial applications [17, 18]. Given their very high cost ($\sim$ \$100.000/m^2), these cells are associated with a low-cost optical concentrator, such as Fresnel lenses, in order to significantly limit the amount of material used. This makes it possible to consider building large-scale photovoltaic power plants while limiting the total cost per unit area. However, high-resolution photovoltaic modules require a two-axis tracking system. Thus, they value mainly under direct illumination and can only be used in the arid and semi-arid regions of the solar belt of the terrestrial globe, where direct illumination is high.

4˙1. *Multi-junction architecture and performance.* – Multi-junction cells consist of a stack of several pn junctions in series. They are composed of semiconductor materials with increasing bandgap widths, each optimized for a different part of the solar spectrum. For a given number of junctions and a given illumination spectrum, there is an

TABLE I. – *Efficiency versus the number of junctions.*

Number of junctions	Efficiency under 1 sun	Maximum efficiency under concentration
1-junction	33.7%	40.8%
2-junction	42.9%	55.7%
3-junction	49.3%	63.8%
∞-junction	68.2%	86.8%

optimal choice of bandgap widths giving the highest efficiency: for example 33.7% for a homojunction, 42.9% for two junctions, 49.3% for three junctions, and 68.2% for an infinite stack of absorbers. This theoretical yield can reach 87% under high concentration (table I). In practice, the greater the number of junctions, the more complex the architecture and technology of the solar cell, and the less the efficiency increases.

The major components of III-V cells are gallium arsenide (GaAs) and indium phosphide (InP), which can be combined with other materials such as aluminum (Al) or tin (Sb) to forming ternary or quaternary alloys such as $Al_xGa_{1-x}As$ or $In_xGa_{1-x}As_yP_{1-y}$. The growth of the layers is carried out by gas phase epitaxy from organometallic compounds. The elements used as dopants "N" are S, Se, Te, Sn while those used as dopants "P" are Zn, Be, Mg, Cd. The elements Si, C, Ge can act as dopants n or dopants p depending on whether they replace, respectively, a gallium (Ga) or arsenic (As) atom in the crystal structure [19, 20].

Figure 12 shows the relationship between the band gap and the crystal lattice parameter of several ternary and quaternary compounds as well as the transitions between them when the relative proportions of the elements in the compound are varied. This

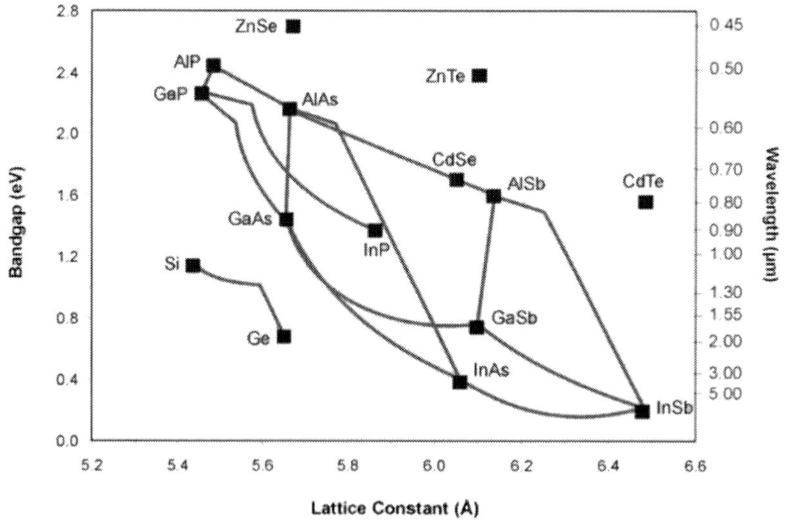

Fig. 12. – Relationship between the bandgap and the lattice constant of several III-V compounds.

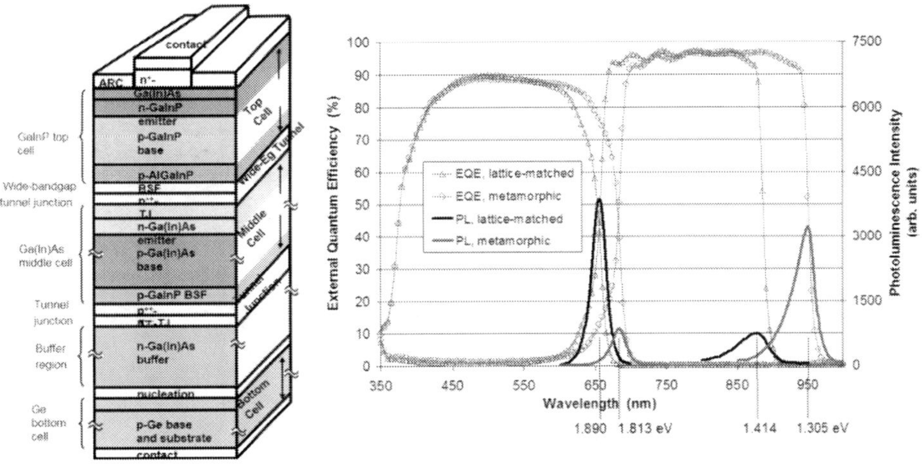

Fig. 13. – Left: scheme of a typical III-V tandem cell structure. Right: external quantum efficiency of a tandem cell.

diagram can be used to identify decreasing bandwidth semiconductors with lattice parameter' sizes (*e.g.*, Ge, GaAs and AlAs have roughly the same lattice parameter with different band gaps). It also makes it possible to manufacture substrates having a gradient of lattice parameter in the case where the successive layers do not have suitable lattice constants (for example for a heterostructure between GaP and $GaAs_xP_{1-x}$). This is made possible by the low rate of growth of epitaxial techniques.

The triple junction cells marketed for terrestrial concentration systems are extremely sophisticated in their construction and use: they can have up to 60 epitaxial layers, 3 junctions and two tunnel diodes. Figure 13 (left) is exhibiting a typical tandem cell structure while fig. 13 (right) is showing the external quantum efficiency as expected from these structures.

As for fabricated cells, the highest efficiencies were obtained with structures based on stacks of III-V compounds, with direct bands and low temperature coefficients, manufactured by epitaxy. A record 46% concentration (508X) was achieved in 2016 in the laboratory (FhgISE, DE) for triple-junction III-V cells.

Due to the series connection, the current produced by the multi-junction cell is limited by the cell producing the lowest photocurrent. The consequence is that these cells are sensitive to changes in the solar spectrum that can occur during a day or a season and between different places.

Several geometries of lenses exist for the concentration purpose: curved cylindrical lenses with linear focus where the engravings are parallel and lenses with circular symmetry, planar or spherical. Planar lenses with circular symmetry are the most commonly used in high concentration (fig. 14(a)). The modules are composed of assembly of lenses and multijunction cells (fig. 14(b)).

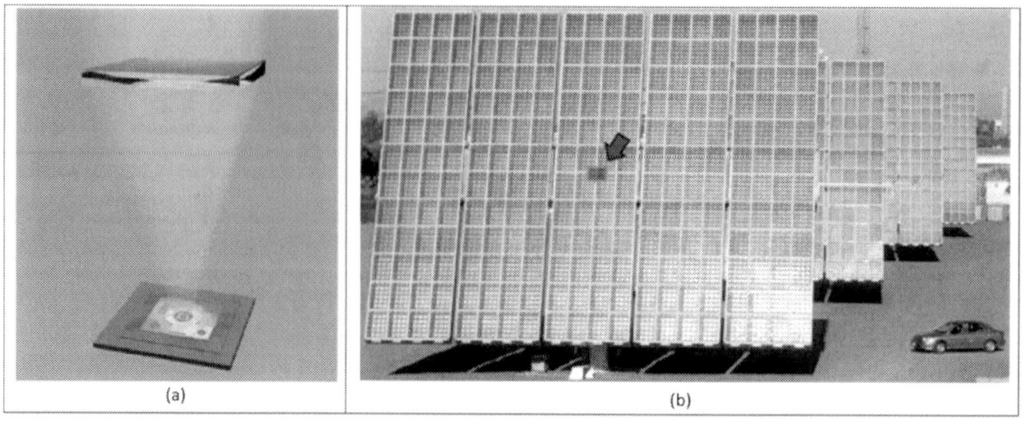

Fig. 14. – (a) Scheme of a planar lense that concentrates light on a multijunction cell. (b) Concentrated photovoltaic modules.

5. – Prospects of medium and long-term technological breakdowns

The technologies presented in the previous paragraphs are all commercial. The focus was on inorganic based photovoltaics using silicon wafers, CIGS or CdTe thin films or multijunction tandem cells. In the medium and long term, other channels are likely to create genuine technological breakthroughs that could lower the costs of the modules. Among the innovations expected, i) advanced architectures allowing to obtain very high efficiencies, thanks to relevant optical managements (down-conversion, up-conversion, plasmonics, ...), ii) thin organic or hybrid films (inorganic and organic combination) which can be manufactured at a very low cost, and iii) the emerging branch of perovskite cells [21-23], which is currently experiencing considerable enthusiasm in the scientific community and has demonstrated more than 20% efficiency in less than 5 years of investigations.

REFERENCES

[1] Shockley William and Queisser Hans J., "Detailed Balance Limit of Efficiency of P-n Junction Solar Cells", *J. Appl. Phys.*, **32** (1961) 510; doi:10.1063/1.1736034.
[2] Hu Chenming and White Richard M., *Solar Cells: From Basics to Advanced Systems* (McGraw-Hill, New York) 1983.
[3] Goetzberger Adolf, *Crystalline Silicon Solar Cells: Technology and Systems Applications* (John Wiley & Sons Ltd) 1998.
[4] Goetzberger Adolf and Volker U. Hoffmann, *Photovoltaic Solar Energy Generation* (Springer, Berlin [u.a.]) 2010.
[5] Green Martin A., "Lambertian Light Trapping in Textured Solar Cells and Light-Emitting Diodes: Analytical Solutions", *Prog. Photovoltaics: Res. Appl.*, **10** (2002) 235; doi:10.1002/pip.404.
[6] Green Martin A., *Third Generation Photovoltaics Advanced Solar Energy Conversion* (Springer, Berlin, New York) 2006.

[7] *Silicon and Ferrosilicon: Global Industry Markets and Outlook*, 14th edition (Roskill) 2014, https://roskill.com/product/silicon-ferrosilicon/.

[8] Solanki C. S., *Solar Photovoltaics: Fundamentals, Technologies and Applications* (PHI Learning Private Limited) 2012.

[9] Markvart T. and Castañer Luis, *Solar Cells Materials, Manufacture and Operation* (Elsevier Science, Oxford) 2005.

[10] Pizzini Sergio, "Towards Solar Grade Silicon: Challenges and Benefits for Low Cost Photovoltaics", *Sol. Energy Mater. Sol. Cells (PVSEC 18)*, **94** (2010) 1528; doi:10.1016/j.solmat.2010.01.016.

[11] Nakajima K. and Noritaka Usami (Editors), *Crystal Growth of Si for Solar Cells, Advances in Materials Research*, Vol. **14** (Springer Verlag, Berlin) 2009.

[12] Poortmans Jef and Arkhipov Vladimir (Editors), *Thin Film Solar Cells: Fabrication, Characterization and Applications, Wiley Series in Materials for Electronic and Optoelectronic Applications* (Wiley, Hoboken, NJ) 2006.

[13] Romeo A., Terheggen M., Abou-Ras D., Bätzner D. L., Haug F.-J., Kälin M., Rudmann D. and Tiwari A. N., "Development of Thin-Film Cu(In,Ga)Se2 and CdTe Solar Cells", *Prog. Photovoltaics: Res. Appl.*, **12** (2004) 93; doi:10.1002/pip.527.

[14] Gorji Nima E., Perez Mauricio D., Reggiani Ugo, and Sandrolini Leonardo, "A New Approach to Valence and Conduction Band Grading in CIGS Thin Film Solar Cells", *Int. J. Eng. Technol.*, **4** (2012) 573; doi:10.7763/IJET.2012.V4.435.

[15] Alonso M. I., Garriga M., Durante Rincón C. A., Hernández E. and León M., "Optical Functions of Chalcopyrite CuGaxIn1-xSe2 Alloys", *Appl. Phys. A*, **74** (2002) 659; doi:10.1007/s003390100931.

[16] Kodigala Subba Ramaiah, *Cu(InGa)Se2 Based Thin Film Solar Cells* (Academic Press, London) 2010, http://www.sciencedirect.com/science/book/9780123736970.

[17] Kurtz S., *Opportunities and Challenges for Development of a Mature Concentrating Photovoltaic Power Industry (revision)* (National Renewable Energy Laboratory (NREL), Golden, CO) 2012, http://www.osti.gov/scitech/biblio/935595.

[18] Luque Antonio and Hegedus Steven, *Handbook of Photovoltaic Science and Engineering*, second edition (Wiley) 2011.

[19] Martí A. and Luque A., *Next Generation Photovoltaics: High Efficiency through Full Spectrum Utilization* (Taylor & Francis) 2003.

[20] Andreev V. M., Grilikhes V. A. and Rumyantev V. D., *Photovoltaic Conversion of Concentrated Sunlight* (Wiley) 1997, http://eu.wiley.com/WileyCDA/WileyTitle/productCd-0471967653.html.

[21] Green Martin A., Ho-Baillie Anita and Snaith Henry J., "The Emergence of Perovskite Solar Cells", *Nat. Photon.*, **8** (2014) 506; doi:10.1038/nphoton.2014.134.

[22] Mingzhen Liu, Johnston Michael B. and Snaith Henry J., "Efficient planar heterojunction perovskite solar cells by vapour deposition", *Nature*, **501** (2013) 395 doi:10.1038/nature12509.

[23] Albrecht Steve, Saliba Michael, Correa Baena Juan Pablo, Lang Felix, Kegelmann Lukas, Mews Mathias and Steier Ludmilla *et al.*, "Monolithic Perovskite/silicon-Heterojunction Tandem Solar Cells Processed at Low Temperature" *Energy Environ. Sci.* **9** (2016) 81; doi:10.1039/C5EE02965A.

Concentrating Solar Power Systems

R. Pitz-Paal(*)

DLR Institute of Solar Research - Linder Höhe, 51147 Cologne, Germany

Summary. — Development of Concentrating Solar Power Systems has started about 40 years ago. A first commercial implementation was performed between 1985 and 1991 in California. However, a drop in gas prices caused a longer period without further deployment. It was overcome in 2007 when new incentive schemes for renewables in Spain and the US enabled a commercial restart. In 2016, almost 100 commercial CSP plants with more than 5 GW are installed worldwide. This paper describes the physical background of CSP technology, its technical characteristics and concepts. Furthermore, it discusses system performances, cost structures and the expected advancement.

1. – Introduction

Concentrating Solar Thermal Power (CSP) systems use high-temperature heat from concentrating solar collectors to generate power in a conventional power cycle instead of —or in addition to— burning fuel. The thermal energy accumulated by the concentrating collectors can be stored at low cost in a thermal energy storage (*e.g.* a tank with molten salt) before it is provided to a conventional power cycle (fig. 1). Thus, electricity production can be shifted to periods of high load demand and the technology is

(*) E-mail: `robert.pitz-paal@dlr.de`

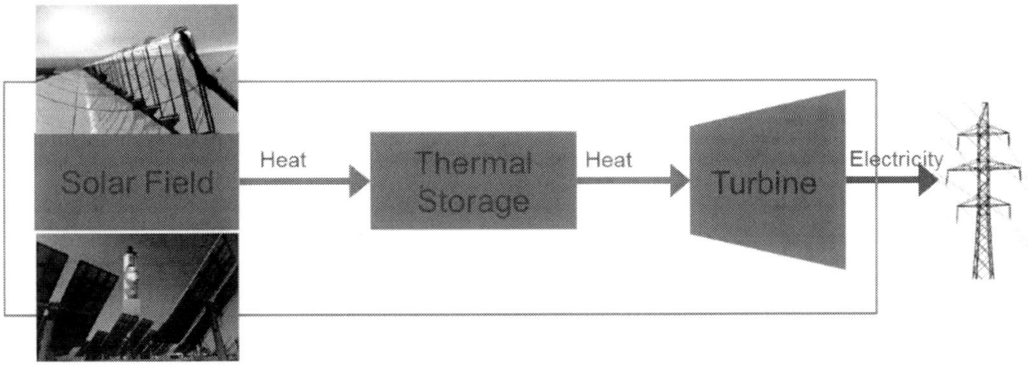

Fig. 1. – The concept of Concentrating Solar Power Systems (CSP). Source DLR.

capable to complement the production from wind and photovoltaic as depicted in fig. 2. Firm supply can be guaranteed if additional fuel (*e.g.* from biomass) is used during the uncommon periods when the thermal storage is empty [1].

2. – Technical background of CSP

2·1. *Solar resource.* – Only direct radiation can be concentrated in optical systems (fig. 3). In order to achieve significant concentration factors, sun-tracking is required, adding complexity to the system. Therefore, CSP is most suitable for centralized power production, where maintenance can be performed efficiently, and in areas with high, direct solar radiation levels. The concentration of sunlight is achieved by mirrors directing the sunlight onto a heat exchanger (receiver/absorber), transferring the absorbed energy to a heat transfer fluid.

2·2. *Concentration of sunlight.* – Radiation energy (unit: Watt) that is emitted from a point-like source is diluted with increasing distance from its source. This means that the energy density E (unit: $\mathrm{W/m^2}$) is reduced with increasing distance, because the emitted

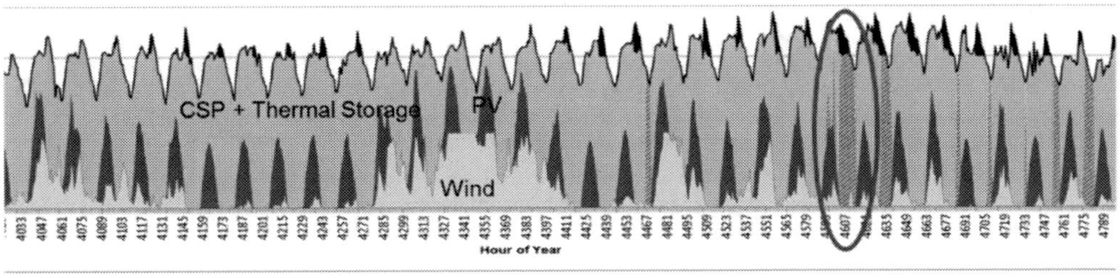

Fig. 2. – Electric load matched by a mix of wind, PV and CSP with thermal storage: During periods indicated in gray co-combustion in CSP systems is required to match the load. Source DLR.

 DOI: 10.1051/epjconf/201714800008

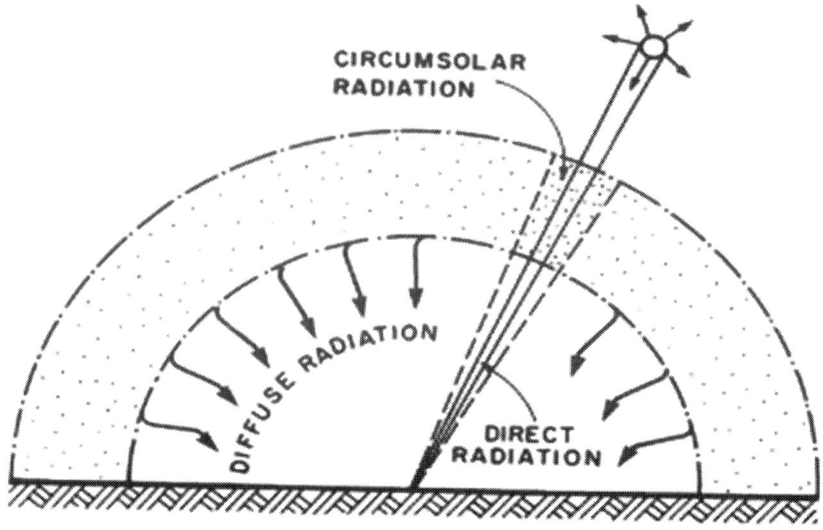

Fig. 3. – Definition of direct, diffuse and circumsolar radiation. Only direct irradicance can be used for CSP [2].

energy is distributed over a larger surface. Concentration of radiations aims to increase the energy density E of the radiant energy for optimized utilization.

A generic concentrator (fig. 4) consists of a concentrator entrance aperture area A (unit: m^2) in which the radiant energy enters and an exit aperture A' from which the radiation energy leaves. Assuming no losses in the concentrator, the law of conservation of energy leads to the concentration factor C. C is the ratio of the outgoing to the incoming energy density and can be described as the inverse ratio of the respective aperture areas:

$$C = E'/E = A/A'. \tag{1}$$

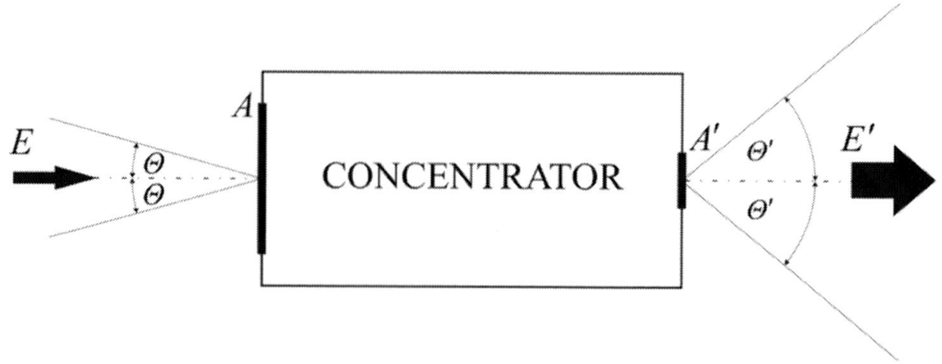

Fig. 4. – Schematic of a solar concentrator [2].

DOI: 10.1051/epjconf/201714800008

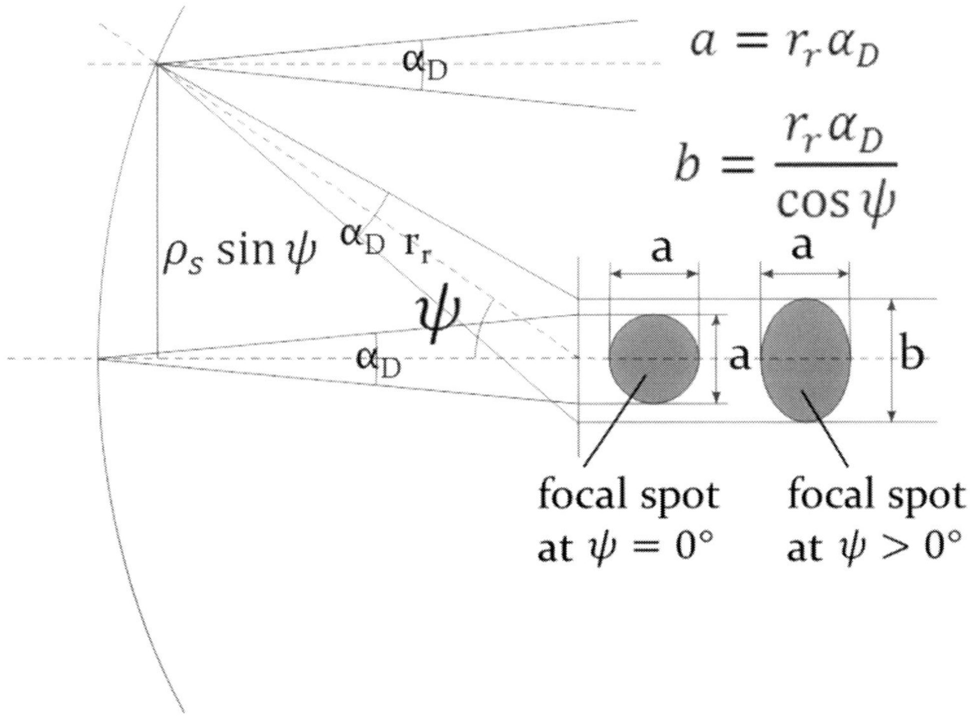

Fig. 5. – Size of the focal spot of an ideal parabola. Source DLR.

As can be derived from complex thermodynamic considerations, ideal concentrators conserve a quantity called entendue. In this case, entendue is the product of aperture area and sinus squared of the radiation cone's half angle θ:

$$A \sin^2 \theta = A' \sin^2 \theta'. \tag{2}$$

From this law, the concentration ratio of an ideal concentrator results as

$$C = A/A' = \sin^2 \theta' / \sin^2 \theta. \tag{3}$$

For a given opening angle θ, the maximum concentration is achieved if $\sin^2 \theta' = 1$ ($\theta' = 90°$).

Thus, for the sun's half angle of 4.653 mrad a maximum concentration ration of $1/\sin^2(4.653\,\text{mrad}) \approx 46200$ can be achieved.

In practice, imaging mirror concentrators rather than lens concentrators are applied as primary concentrators for solar high-temperature applications. Imaging mirror concentrators have better outdoor durability and lower specific costs. Their design generally approximates the parabola shape in a continuous or segmented way (fig. 5). The image of the sun as generated by such a mirror reflector is blurred by the optical imperfections of the parabola concentrator concept itself (aberration) and by imperfect surface characteristics. The resulting image consists of superimposed images of individual concentrator segments having elliptical shapes with diameters depending on the opening angle of the sun's disk α_D, the distance between reflector and focal point r_r, and on the opening angle of the parabola Ψ.

The concentration factor of the parabola is determined by the largest of all superimposed images A_{im} that comes from the rim of the parabola:

$$C = \frac{A_{ap}}{A_{im}}, \tag{4}$$

with

$$A_{ap} = \pi r_r^2 \sin^2\psi \qquad \text{and} \qquad A_{im} = \frac{\pi}{4}\frac{r_r^2\alpha_D^2}{\cos^2\psi},$$

leads to

$$C = \frac{4}{\alpha_D^2}\sin^2\psi\cos^2\psi = 46200\sin^2\psi\cos^2\psi. \tag{5}$$

The maximum concentration achievable by an ideal parabola does not reach the theoretical limit of 46200, but only 1/4 of it at an opening angle of the parabola Ψ of 45°:

$$C_{\max} = 46200 \cdot 0.5 \cdot 0.5 = 11550. \tag{6}$$

The superposition of the images can often be well approximated by a Gaussian distribution. Thus, the energy density at the exit aperture of a parabola concentrator is not constant, but varies from a peak value to zero at an infinite distance. For practical application it is reasonable to use only the central part of the Gaussian profile and discard the rest. The amount of discarded energy is often an outcome of an economic optimization process with typical values ranging from 5% to 10%.

2·3. *Conversion of concentrated sunlight to heat.* – The device used in high-temperature solar concentrators for the conversion of concentrated solar radiation to heat is called receiver. It is designed to absorb the concentrated solar radiation and to transfer as much energy as possible to a heat transfer fluid. Losses have several origins: The absorbing surface may not be completely black, the surface emits thermal radiation to the environment (due to its elevated temperature) and convection as well as conduction occur. The thermal efficiency of the receiver η_{th} is defined by the ratio of the useable heat per second to the incoming solar radiation in the aperture. As shown in fig. 6, η_{th} is decreasing with increasing temperatures due to higher heat losses. On the other hand, higher concentration factors lead to higher efficiencies and both convection and conduction losses are of minor importance at high concentration factors.

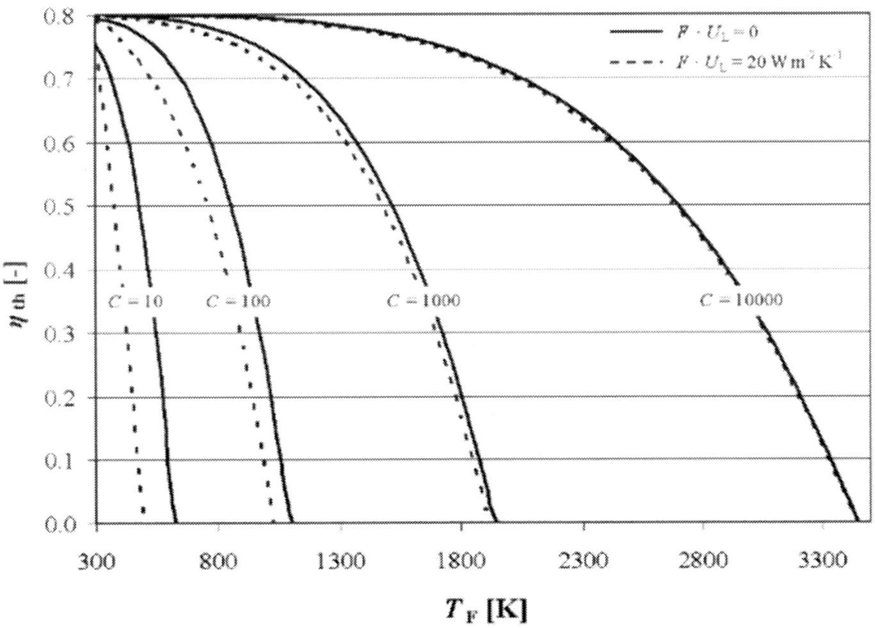

Fig. 6. – Efficiency of a solar receiver as a function of the fluid temperature, the concentration ratio C and the effective convective heat loss coefficient $F\,U_L$ [2].

2·4. *Thermodynamic cycles.* – In high-temperature solar concentrator systems, the absorbed energy is often converted into mechanical energy and subsequently to electricity in a power cycle. A part of that cycle is an engine that continuously transfers heat to mechanical energy. Steam or gas turbine cycles are common examples used in fossil power plants to generate electricity from the heat input provided by the combustion of fuels (fig. 7(a)). However, thermodynamic principles forbid that all heat input is completely transferred to electricity, because a fraction of it must be removed at a lower temperature level.

In each generic power cycle, a working fluid undergoes a number of changes in state. In order to force the system from one state to the next, heat or work must be delivered to or extracted from the fluid. Depending on the kind of working fluid and the state changes, many different power cycles can be set up. An example is the Carnot cycle, an idealized reversible process that consists of two isothermal ($2 \to 3$ and $4 \to 1$) and two isentropic ($1 \to 2$ and $3 \to 4$) changes in state. In fig. 7(b), the corresponding T/S-representation is shown, where S is the specific entropy of the fluid.

The fluid receives heat at the upper temperature T_{H} and at the lower temperature T_0 heat is released. The cycle efficiency η_{C} describing the fraction of heat input Q_{in} that is converted to work W is independent of the working fluid and can be calculated to be

$$\eta_{\mathrm{C}} = \frac{W}{Q_{\mathrm{in}}} = 1 - \frac{T_0}{T_{\mathrm{H}}}\,. \tag{7}$$

Today, there are no power cycles specifically developed for high temperature solar concen-

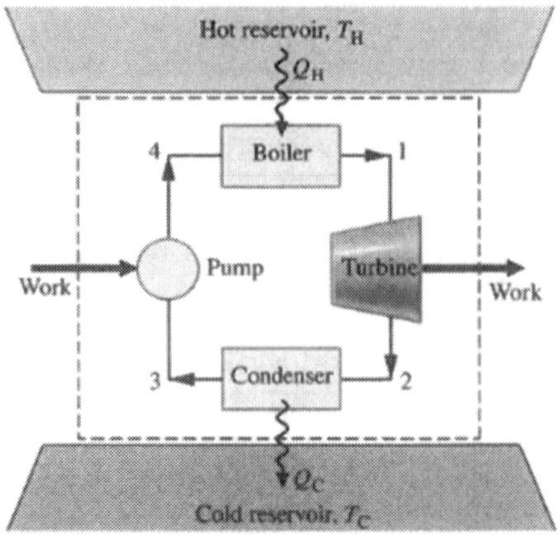

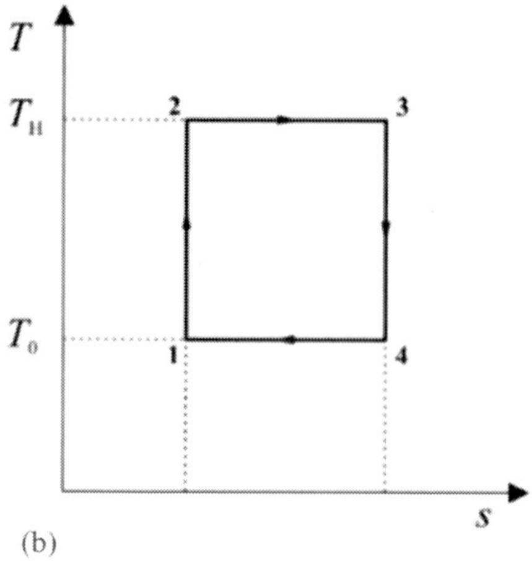

Fig. 7. – Schematic of an ideal Carnot Cycle (a) and state changes in the *T*-*S* representation (b). Source DLR.

trating systems but conventional fossil fuel driven power generation systems are adapted to solar applications. The most relevant ones are steam turbine cycles, gas turbine cycles and Stirling engines. Currently, steam cycles are the most common choice in commercial CSP projects. They are suited to power levels beyond 10 MW and temperatures of up to 600 °C. Steam cycles can be coupled to parabolic trough, linear Fresnel and central receiver systems. Stirling engines are used for small power levels (up to some 10 kW_{el}). Gas turbines offer the potential to exploit higher temperatures than steam cycles (up to 1 200 °C) and cover a wide range of capacities from some hundred kW_{el} to some 10 MW_{el}. At high power levels, gas turbine cycles may be combined with steam cycles to highly efficient combined cycle systems promising to produce the same power output with a 25% smaller solar collector field. Up to now, solar gas turbines have been used in experimental facilities only.

2·5. *Thermal energy storage.* – Like domestic hot water systems, CSP systems have the important advantage of optional thermal energy storage systems (*e.g.* tanks with molten salt), allowing the operation of the plant to continue during transient clouds or after sunset. Thereby, a predictable power supply to the electricity grid can be achieved. In contrast to other renewable systems with electric storage, for which the inclusion of storage capacity always leads to far higher investments and higher electricity prices, CSP systems with storage are potentially cheaper than CSP systems without storage. This becomes clear when comparing a solar power plant without storage of *e.g.* 100 MW_{el} capacity, operated approx. 2000 equivalent full-load hours per year at a typical site, to a system with half the capacity (50 MW_{el}) but the same size solar field and a suitable thermal energy storage (fig. 8). In this case, the smaller power block is used for 4 000 equivalent full load hours so that both systems can produce the same amount of electricity per year. Assuming low storage costs, the investment in the second system could potentially be lower than in the first.

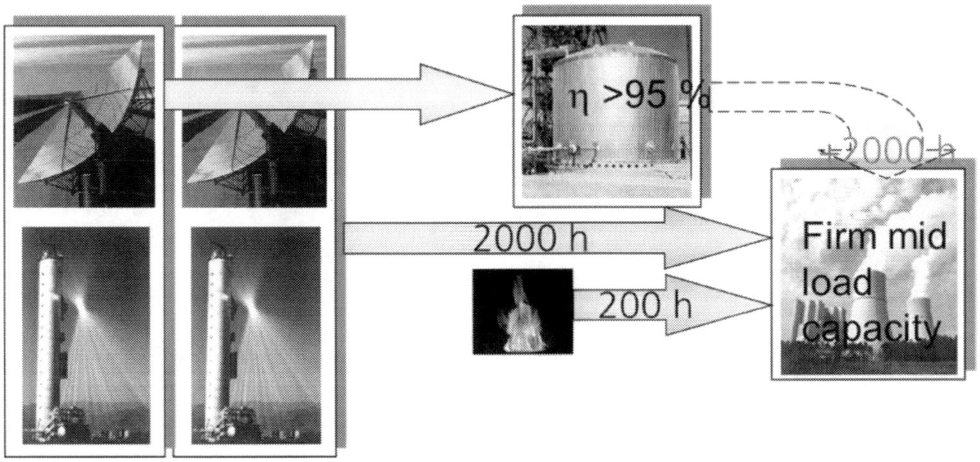

Fig. 8. – Integration of a thermal storage in a CSP system may lead to lower specific electricity cost (see text). Source DLR.

2·6. *Optimizing CSP-systems.* – According to the principles of thermodynamics, power cycles convert heat to mechanical energy more efficiently at higher temperatures. However, the collector efficiency drops at higher absorber temperatures due to higher heat losses. Consequently, for any given concentration factor there is an optimum opera-

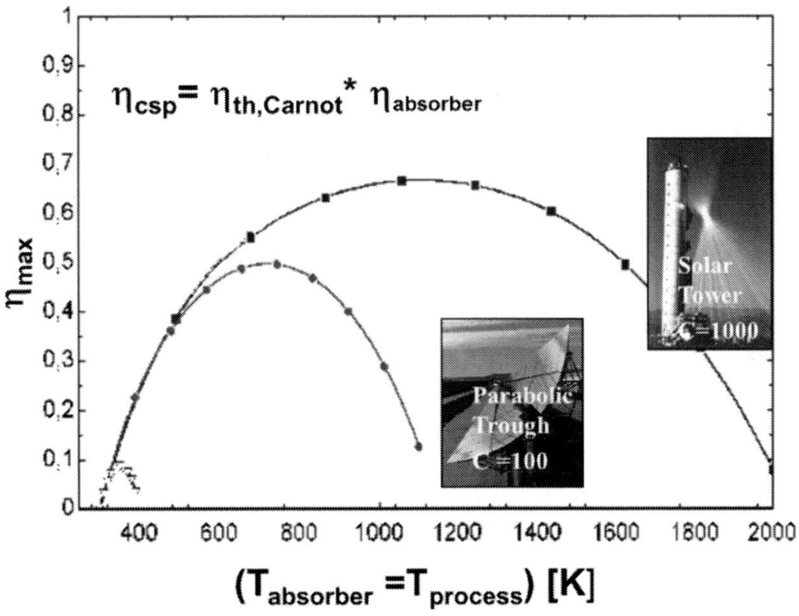

Fig. 9. – System efficiency as a function of upper fluid temperature for different solar concentrating systems. Source DLR.

Fig. 10. – Schematic of a parabolic trough collector (left), linear Fresnel (center) and solar tower system (left). Source DLR.

tion temperature at which the highest conversion efficiency from solar energy to work is achieved. With rising concentration higher optimum efficiencies are achievable. Figure 9 illustrates this characteristic assuming an ideal solar concentrator combined with an ideal Carnot power cycle. If the spectral absorption characteristics of the absorber are perfectly tailored to maximise absorption in the solar spectrum, but avoid thermal radiation losses in the infrared part of the spectrum (selective absorber), additional efficiency gains can be expected in particular at lower concentration factors.

In practice, the optimum operation temperatures will be lower than these theoretical figures, because power cycles with Carnot performance and ideal absorbers do not exist. Furthermore, the impact of frequent operation under part-load conditions throughout the year on the efficiency of the system has to be considered.

3. – Technical concepts

CSP systems can be distinguished by the arrangement of their concentrator mirrors: Line focusing systems like parabolic troughs or linear Fresnel systems (fig. 10, left and middle) only require single axis tracking in order to concentrate the solar radiation onto an absorber tube. Concentration factors of up to 100 can be achieved in practice. Point focusing systems like central receiver systems (fig. 10 right) can achieve concentration factors of several 1 000 at the expense of two-axis tracking. They use a large number of individually tracking heliostats to concentrate the solar radiation onto a receiver located on the top of a central tower.

3·1. *Line focusing systems.* – Most of today's commercial CSP power plants ($\Sigma > 3\,\mathrm{GW_e}$) are based on the design of the ANDASOL parabolic trough power plant (fig. 11)

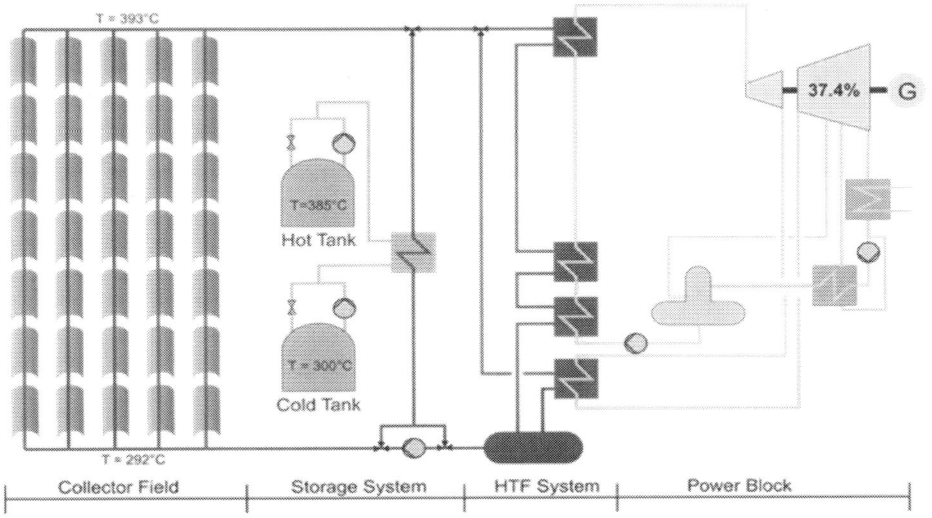

Fig. 11. – Schematic of a system based on parabolic trough collector using thermal oil as heat transfer fluid and 2 tank molten salt storage system [3].

put in operation in Spain in November 2008. The plant design integrates three different fluid cycles: a eutectic mixture of 73.5% diphenyl oxide and 26.5% biphenyl is used as heat transfer fluid in the collector field. It is operated between 290° and 390 °C (the upper temperature limit of this thermal oil) and provides the collected energy through heat exchangers either to the thermal storage cycle or to the power cycle. The power cycle uses water steam like in conventional power plants whereas a eutectic mixture of KNO_3 and $NaNO_3$ acts as a storage medium at almost the same temperature range as the collector fluid. The mixture is chosen as it is about 4 times cheaper than the thermal oil. The salt is contained in two tanks, one on a lower temperature of about 290 °C and, after being heated up in the heat exchanger, in a hot tank at almost 390 °C. The salt mixture material is available extensively in nature and used also as fertilizer with a capacity of many 100 of thousands tons per year. As it freezes at 238 °C, it needs to be kept significantly above that temperature during the lifetime of the power plant. The storage density of such a system is approximately $78\,\mathrm{kWh_{th}/m^3}$ or $42\,\mathrm{kWh_{th}/t}$. To provide a $100\,\mathrm{MW_e}$ trough power plant with a 6 hour of storage capacity about 38 000 t of salt are required. The largest commercial installation that uses such a storage system was put in operation in October 2013. It is the $280\,\mathrm{MW_e}$ Solana plant in Arizona (US) equipped with a 6 hours storage equivalent to $4\,400\,\mathrm{MWh_{th}}$ of storage capacity distributed in 6 pairs of tanks of hot and cold salt each about 38 m in diameter and 14 m high. The heat losses of such tanks are significantly below 3% of the energy throughput and even two weeks of interruption does not require active heating for freezing protection of the storage tanks. Heat exchangers, pumps valves and pipes need to be equipped with electrical trace heating to avoid freezing problems during start-up. Such a power plant concept can achieve annual net solar-to-electric efficiency figures of up to 15%.

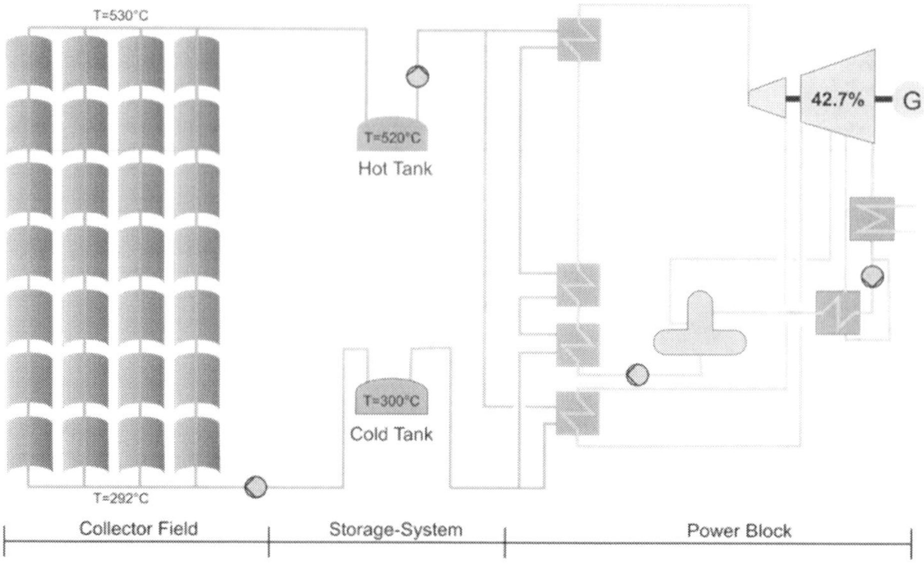

Fig. 12. – Schematic of a system based on parabolic trough collector using molten salt as heat transfer fluid and a 2 tank molten salt storage system [3].

This concept has a number of limitations in terms of power plant efficiency and storage density that led to the development of alternative designs. First, the synthetic oil heat transfer is rather expensive and limited to approximately 400 °C. This confines the potential life steam conditions in the power block and thus the efficiency of the power cycle to values below those of fossil fuel driven steam power plants. Moreover, it limits the amount of heat that can be stored in the salt as it depends on the temperature difference between hot and cold tank. Steam, also used in the power cycle, has been investigated as an alternative heat transfer fluid. However, the integration into a storage system is difficult as the phase change characteristics of the steam during evaporation leads to a temperature profile that does not correspond to the sensible heating of the salt mixture. Using salt not only as a storage medium, but also as heat transfer fluid may overcome many of the limitations as the operation temperature of the salt can be extended to 550 °C which is a more typical operation temperature of conventional steam cycle power plants. Such a replacement would also avoid the expensive thermal oil cycle. The challenge, however, is to avoid freezing of the salt in the extended collector field under any circumstances.

If such a concept is applied to a parabolic trough system (fig. 12) the molten salt needs to be circulated through the whole collector field (> 100 km of piping) and a complete draining of the systems (*e.g.* overnight) does not seem to be an attractive option for freezing protection. In order to prevent the salt from freezing in the absorber tubes, a continuous recirculation, using salt from the cold tank and redirecting it back to the cold tank, poses a solution. Any drop in temperature of the cold tank will be compensated

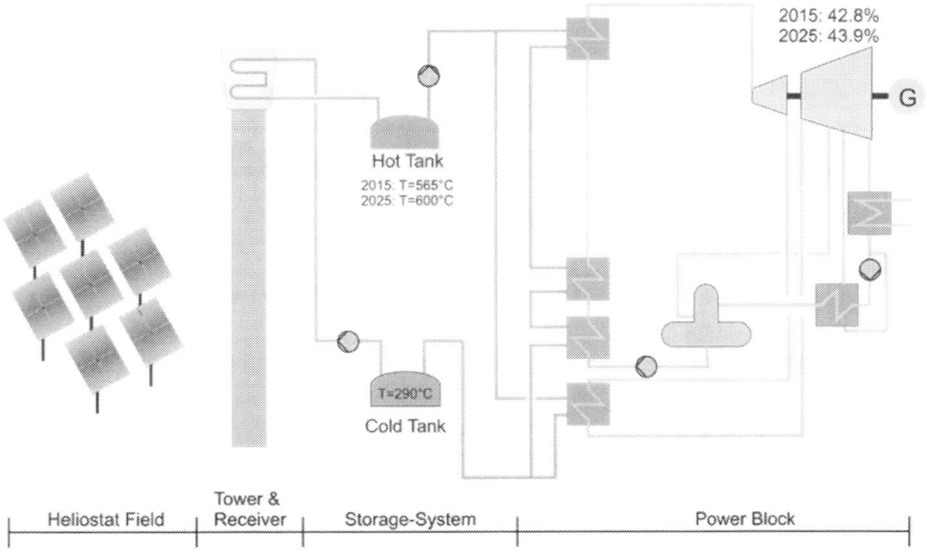

Fig. 13. – Schematic of a solar tower system using molten salt as heat transfer fluid and a two-tank molten salt storage system [3].

by small amounts of solar-heated molten salt from the hot tank. Once the hot tank is empty the heat will be provided by the auxiliary heater. Typically the auxiliary heater is not used after a sunny day. As molten salt always circulates through the solar field, it requires energy for pumping. The additionally installed electric heating devices in the solar field are only installed for emergencies and will not be used in a regular operation of the plant.

3·2. *Solar tower systems.* – In solar tower systems, a field of individually tracked mirrors (heliostats) concentrates the solar radiation on a heat exchanger located on top of a central tower (fig. 10 right). In such a point-focussing system, higher-concentration factors ($\approx 1\,000$) can be archived compared to parabolic trough technology (≈ 100) that generates a line-focus. The collection of the energy is not performed by an extended piping circuit of more than 100 km length (including the parabolic trough receiver tubes) like in a parabolic trough field but rather by transferring the radiation energy as light to a central location. This limits the piping length to a couple of 100 m necessary to transport the heat transfer fluid from the top of the tower to the storage and steam generator system located on the ground (fig. 13). This configuration has the benefit that gravity supports the draining of the complete salt fluid back to the storage tanks during times when operation is not scheduled (*e.g.* overnight). This reduces the auxiliary electric power requirements to heat the pipes in order to avoid freezing, in the case that molten salt is used as heat transfer and storage fluid. However, such a design also shows some significant drawbacks: A liquid salt column of 100 m equals 19 bar pressure. As the tower in a commercial solar system is typically more than 150 m tall and the pressure

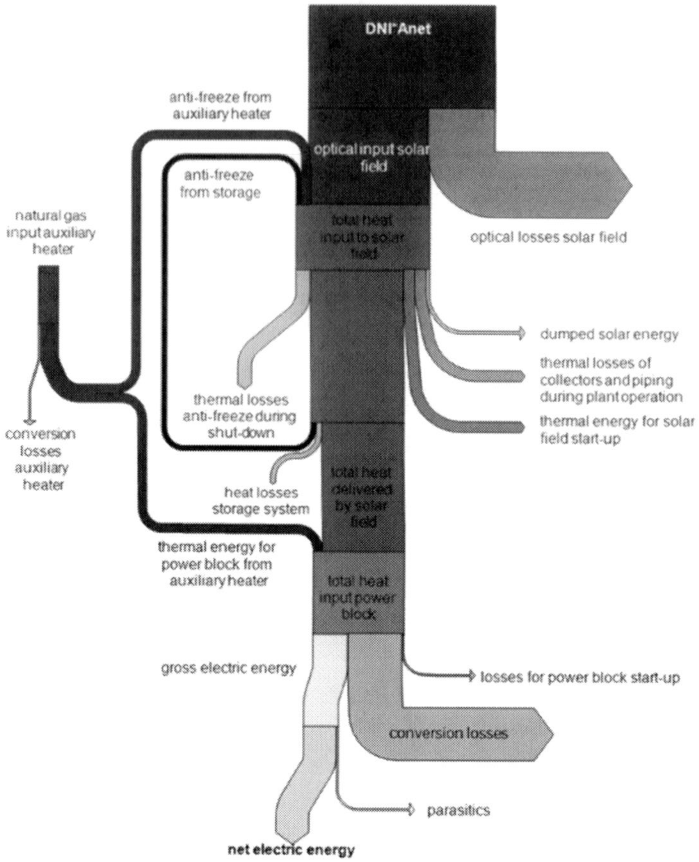

Fig. 14. – Sankey diagram for energy flows in a molten salt line focusing system [4].

in the storage tanks at the ground level should be kept close to atmospheric pressure to limit construction costs, the salt circulation system cannot be designed as a closed circuit in which only friction losses need to be overcome by the pump. In an open cycle, the energy required to lift the salt to the tower top needs to be invested.

4. – Performance

In fig. 14, the annual energy flow in a CSP line focusing system is depicted qualitatively. Energy flows in tower systems differ only slightly. Optical losses associated with the solar field are caused by geometrical imperfections, surface impairments (*e.g.* dust) and extinction (aerosols). Thermal losses are linked to heat losses of high temperature energy flows and to auxiliary warming-up effects. Power block losses are due to thermodynamic constraints that require heat removal of the cycle at low temperatures.

Performance and cost figures of CSP systems depend strongly on the power plant designs with respect to its size, its capacity factor (storage) as well as plant location

TABLE I. – *Data for reference system concepts to evaluate performances and costs for different site conditions* [4].

Design parameters in 2015	Unit	Parabolic Trough	Solar Tower
Site		Ouarzazate, Morocco	
Direct normal irradiation (DNI)	[kWh/(m^2 a)]	2017/2558/2935	
Solar collector/heliostat		Ultimate Trough®	Stellio®
Heat transfer fluid (HTF)		thermal oil	molten salt
Storage medium		molten salt	molten salt
Maximum HTF temperature	[°C]	393	565
Thermal energy storage capacity (full load hours)	[h]	7.5	9[(a)]
Gross electrical output	[MW]	160	150

[(a)] Result of LCOE optimisation.

and solar resource. In order to discuss this effect, the boundary conditions of different solar reference plant concepts investigated by IRENA [4] are summarized in table I. The concepts comprise of trough and tower technology with a certain capacity and storage size investigated for three different solar conditions that are representative for typical CSP markets.

Performance calculations of those concepts have been conducted with the numerical tool Greenius ® [4] and are shown for trough systems in table II and for tower systems in table III. Advancements in technology until 2025 are considered (*e.g.* replacing thermal oil by molten salt in parabolic trough in 2025) by different sets of technical parameters. The capacity factor shown here is defined as the annual electricity energy provided by the power plant divided by the nominal capacity multiplied by 8760 hours. Tower systems are expected to have slightly higher efficiencies compared to parabolic trough systems and are capable to achieve higher capacity factors.

TABLE II. – *Electric production and system performance figures of a 160 MW_e parabolic trough system for state of the art technology (thermal oil heat transfer fluid) and estimations for 2025 (molten slat heat transfer fluid)* [4].

	Net Electrical Output [GWh/a]			Net Plant Efficiency [%]		Capacity Factor [%]	
	2015	2025	Variation	2015	2025	2015	2025
DNI 2000	526.6	551.3	+4.7%	15.2%	17.5%	37.6%	39.3%
DNI 2550	575.9	624.6	+8.4%	14.9%	16.6%	41.1%	44.6%
DNI 2900	639.7	683.6	+6.9%	15.1%	16.9%	45.6%	48.8%

TABLE III. – *Electric production and system performance figures of a solar tower reference system for state of the art technology and estimations for 2025* [4].

	Net Electrical Output [GWh/a]			Net Plant Efficiency [%]		Capacity Factor [%]	
	2015	2025	Variation	2015	2025	2015	2025
DNI 2000	574.0	610.0	+6.2%	14.4%	17.1%	43.7%	46.4%
DNI 2550	598.4	644.0	+7.6%	15.5%	18.3%	45.5%	49.0%
DNI 2900	656.6	684.4	+4.2%	15.6%	18.2%	50.0%	52.1%

5. – Cost

Based on information of commercially built power plants on investments of the different components and subsystems as well as on typical surcharges for engineering, procurement, construction (EPC), and owner's cost for the project development, land preparation and other factors, investments cost for the defined trough and tower reference concepts were calculated as shown on the left in fig. 15 and fig. 16. Technology advancements and other factors that will be discussed later are expected to lead to a cost reduction until 2025 as presented on the right hand side of the figures mentioned. The solar concentrator (that consists of a solar field in the case of the parabolic trough and of a heliostat field, a tower and a receiver for the tower system) has the major cost share followed by the power block. Today, high surcharges in EPC and owners cost exist due the early market situation of the technology.

In order to compare the technologies, levelised cost of electricity (LCOE) are calcu-

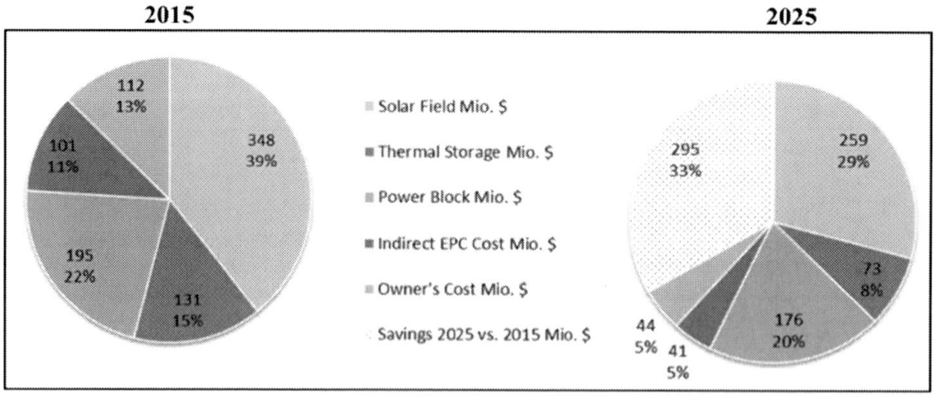

Fig. 15. – Comparison of CAPEX structure in 2015 and 2025 for 160 MW parabolic trough power plant in Mio US$ [4].

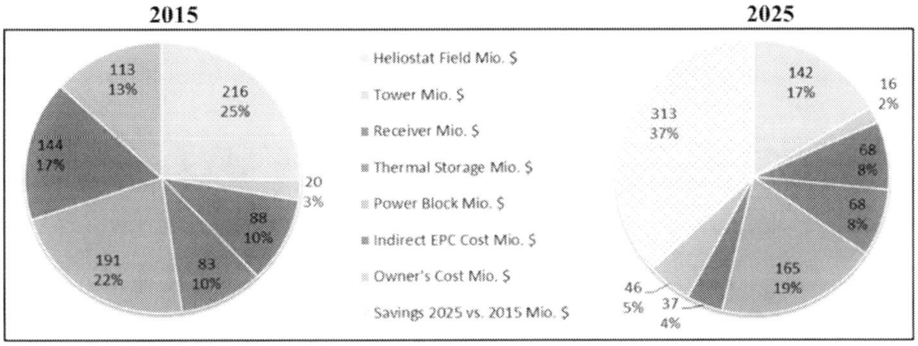

Fig. 16. – Comparison of CAPEX structure in 2015 and 2025 for 150 MW solar tower power plant in Mio US$ [4].

lated using the following equation:

$$(8) \qquad LCOE = \frac{C_{invest} \cdot (f_{annuity} + f_{ins.,ann.}) + C_{O\&M,ann.} + C_{elec} + C_{fuel}}{E_{el,net,ann.} \cdot f_{avail,plant}},$$

with C_{invest} being the capital investment, $C_{O\&M,ann}$ the annual operation and maintenance costs, C_{elec} the annual electricity costs (if any), C_{fuel} the annual fuel costs (if any), $E_{el,net,ann}$ the annual net electricity production, $f_{annuity}$ the weighted average cost of capital factor, $f_{ins,ann}$ the fraction of capital costs needed to cover the annual insurance costs and $f_{avail,plant}$ as the average annual plant availability.

The LCOE in US $-ct/kWh is compared in fig. 17 between trough and tower systems for the year 2015 and 2025. The colored margin shows the impact of a variation in the solar resource and in different financing conditions expressed by the factor weighted average cost of capital (WACC $= f_{annuity} + f_{ins,ann}$). In addition, electricity prices from power purchase agreements announced in 2015 of a trough and a tower plant to be built in Morocco at around 2 500 kWh/m^2 a (Noor II &III) are indicated in the picture. They are below the average expected value of the model calculation because financing costs of those projects where lower than assumed in the model calculations due to the availability of soft loans provided by development banks.

Cost of troughs and towers are very similar and are expected to be reduced significantly over the next decade. Trough technology may keep up with the tower system if it is possible to replace the thermal oil heat transfer fluid by molten salt as it is state of the art in tower technology already today. CSP electricity prices range today between 12 and 20 US-cts/kWh depending on financing and site conditions. At this level the technology is not cost competitive with conventional fossil fuel based options. If a level below 10 US-cts/kWh will be reached by 2025, first niche markets where CSP complements the electricity portfolio of wind and PV systems may become available. However, this also depends on the cost reduction of other electric storage technologies like lithium-ion batteries.

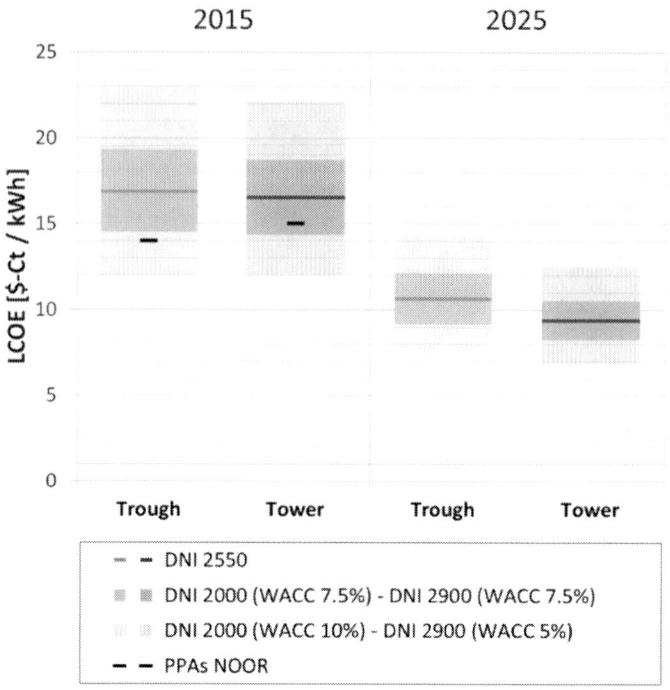

Fig. 17. – LCOE in 2015 and in 2025 for parabolic trough and tower technology, different solar resource and financing conditions [4].

5·1. *Cost drivers.* – As shown in the previous section, both parabolic trough and tower technology offer significant cost reduction potentials. However, the cost drivers for both technologies are quite different.

In general, LCOE reductions can be achieved either by improving the efficiency of the plant resulting in a higher electricity output or by reducing the costs (CAPEX and/or OPEX). In this study, performance improvements as well as cost reductions contribute to the expected LCOE reductions until 2025. Figure 18 gives an overview of the main cost drivers for both technologies.

For trough technology, two major cost drivers can be identified. The increase of the temperature level enabled by the switchover to molten salt as HTF leads to a higher cycle efficiency of the power block and lower investment costs for the thermal storage due to the size reduction by more than 50% and the omission of a heat exchanger. Both effects together account for 1.9 $-ct/kWh of LCOE reduction. The second important cost driver is the reduction of the solar field costs which is closely related to the usage of a trough collector with wider aperture and thus less collector units. Together both factors lead to LCOE reductions by 1.75 $-ct/kWh. The residual LCOE reduction is mainly achieved by the reduction of indirect EPC and owner's cost which are assumed to reduce proportional to the direct EPC cost.

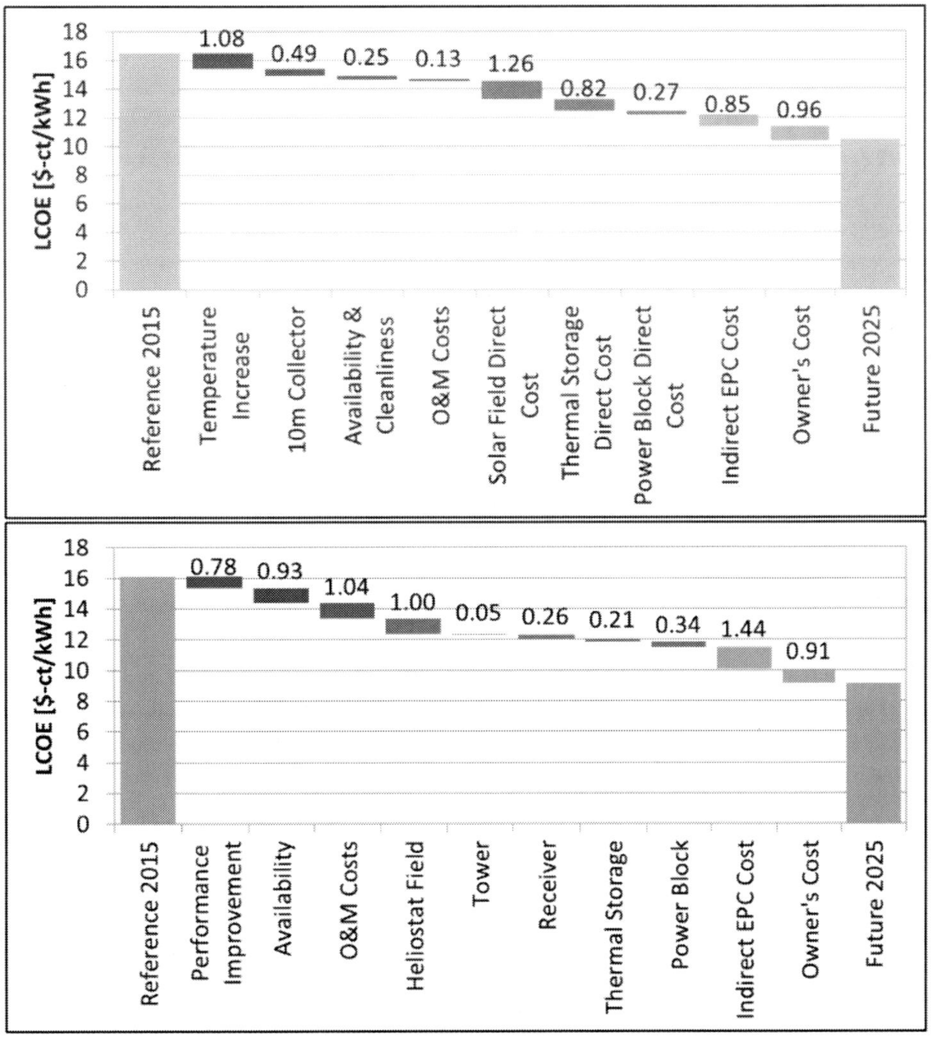

Fig. 18. – Contributing factors to the LCOE reduction from 2015 to 2025 for trough (top) and tower (bottom) systems: Performance improvement (violet), OPEX reduction (red), direct CAPEX reduction (dark green) and indirect CAPEX reduction (light green) [4].

For tower technology, the major cost driver is the gain of experience. Since many of the EPC contractors and project developers built their first tower projects in 2015, risk margins are still high, the commissioning phase is extended and the operational costs are higher (at least in the first years of operation) compared to mature trough technology. The lack of experience in 2015 is represented by three factors that are altered compared to the values for troughs. The plant availability is assumed to be only 0.93 in 2015 instead of 0.99 in order to account for longer outages due to unscheduled additional maintenance and replacement of broken components. Extended commissioning phases with replacement of broken components could be seen for several tower projects built until now. These additional maintenance and commissioning efforts are also responsible for the assumption of increased O&M costs (2% of CAPEX in 2015 instead of 1% in 2025). Indirect EPC costs are expected to be 9%-points higher in 2015 due to additional risk surcharges within the supply chain. The mentioned three parameters are only altered for the year 2015. In 2025 the percentages for all indirect costs and availability are the same as for troughs which reduces the LCOE by 3.4 \$-ct/kWh. With the exception of the heliostat field, the reduction of direct EPC costs is marginal. Performance improvements excluding availability have about half as much influence for tower technology (0.78 \$-ct/kWh) as for trough technology (1.08+0.49=1.57 \$-ct/kWh).

6. – Summary and conclusion

In solar thermal power systems, concentrating solar collectors provide high temperatures to a power engine to generate mechanical energy that is converted to electricity using a generator. It benefits from a mature power plant technology that has been optimized for almost a century using heat based on fossil or nuclear resources. Today, several options to concentrate direct sunlight are pursued. In linear concentrators like the parabolic trough or linear Fresnel systems, a line focus is achieved by one axis tracking of 1d curved reflector segments. In point focusing systems like towers systems, a point focus is generated by 2d curved reflector segments and two axis tracking. All options can be combined with thermal energy storage to provide energy on demand. Thus, technology concepts that are based on molten salt both as a storage and as a heat transfer fluid are in the focus of the current deployment. A significant cost reduction potential provides the basis allowing CSP power plants to achieve competitiveness in the next 10 years. Trough and tower technologies bear advantages and drawbacks that do not allow a clear preference for one technology today.

REFERENCES

The work has been compiled partly from previous publications of the author:

[1] Pitz-Paal Robert, Buck Reiner, Heller Peter, Hirsch Tobias and Steinmann Wolf-Dieter, *Solar Thermal Power Production*, in *Transition to Renewable Energy Systems* (Willey-VCH) 2013, pp. 307–338.

[2] Pitz-Paal Robert, *High Temperature Solar Concentrators*, in *Solar Energy Conversion and Photoenergy Systems*, edited by Blanco Galvez Julian and Malato Rodriguez Sixto (Eolss Publishers, Oxford, UK), 2007.

[3] Pitz-Paal Robert, Giuliano Stefano and Wittmann Michael, *Concepts for Cost Reduction in CSP Power Plants*, in *Energie Erzeugung, Netze, Nutzung - Vorträge auf der DPG Frühjahrstagung, 16–18.3.2015, Berlin* (Deutsche Physikalische Gesellschaft (DPG) Frühjahrstagung) 2015, pp. 42–52.

[4] Dieckmann Simon, Dersch Jürgen, Lüpfert Eckhard, Pitz-Paal Robert, Giuliano Stefano, Puppe Michael, *Assessment of the Cost Reduction Potential for Concentrating Solar Power Until 2025*, study performed by the German Aerospace Center (DLR) for the International Renewable Energy Agency (IRENA) 2016.

 DOI: 10.1051/epjconf/201714800009

Concentrated Solar Power: Components and materials

A. Kribus

School of Mechanical Engineering, Tel Aviv University - Tel Aviv 69978, Israel

Summary. — CSP technologies are well developed and offer many advantages compared to other renewable energy options. They can also be very effective in many locations with high solar radiation around the world. However today they are less competitive than other technologies. Understanding the limitations, and identifying opportunities for improvements, requires a detailed analysis of the energy conversion processes, the needed components, and the required technologies for these plant components. Here we present the three main energy conversion steps in a CSP plant, the behavior and limitations of the technologies that are currently used in commercial CSP plants, and some directions for development of plant components that will offer better performance.

1. – Energy conversion efficiency

The technological and commercial success of CSP technologies depends on many issues, including the conversion efficiency, the annual amount of produced electrical energy, the cost to manufacture, install and operate the plants, reliability of operation, etc. This article will focus on the aspect of conversion efficiency and how it affects the selection of materials and technologies that constitute a CSP plant. For an overview of CSP systems see the article "Concentrated solar power: systems" by Robert Pitz-Paal.

 DOI: 10.1051/epjconf/201714800009

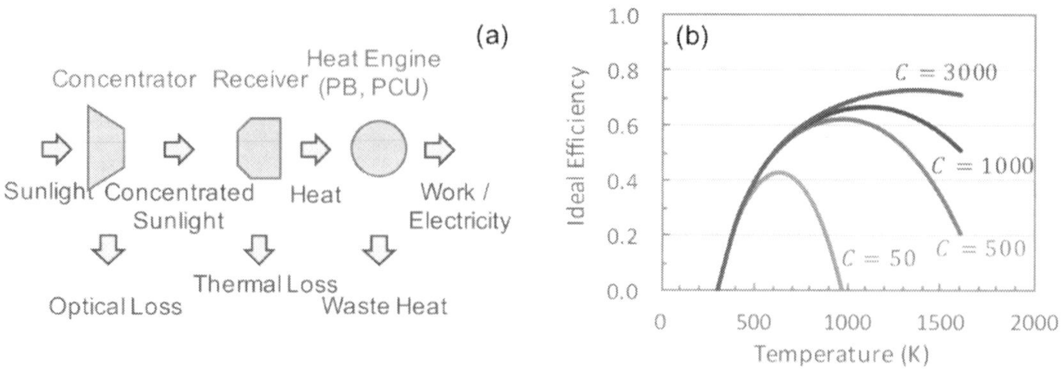

Fig. 1. – (a) Schematic of main energy conversion steps. (b) Ideal plant efficiency as a function of concentration ratio and receiver temperature.

The overall plant efficiency η can be expressed as a product of several subsystem efficiencies, corresponding to the sequence of steps that occur during the process of converting sunlight to electricity, as shown in fig. 1(a):

$$\eta(C, T_{rec}) = \eta_{opt}(C) \cdot \eta_{rec}(C, T_{rec}) \cdot \eta_{PB}(T_{rec}). \tag{1}$$

η_{opt} is the optical efficiency: the fraction of the solar radiation collected by the concentrator field that actually enters the receiver aperture. η_{rec} is the receiver efficiency, the fraction of radiation power entering the receiver that exits as thermal energy in the fluid stream. η_{PB} is the power block efficiency, which represents actually two conversions: the conversion of thermal energy in the fluid to mechanical energy by the heat engine, and conversion of the mechanical energy to electrical energy by the generator. These subsystem efficiencies depend on the concentration ratio C and on the receiver temperature T_{rec}. For simplicity, we assume that the receiver temperature is uniform (although in real receivers there is a temperature distribution and this affects their efficiency), and that the fluid provided by the receiver to the power block is at the receiver temperature (again not always accurate in realistic systems).

As a first step we can estimate the upper limit on plant efficiency, which would be obtained if all components are ideal and all losses are reduced to the lowest possible value. The optical efficiency is in general a complicated function of the concentrator geometry and material, and it can change in time depending on the apparent motion of the sun. However, an ideal optical system can in principle reach $\eta_{opt} = 1$. The receiver efficiency, on the other hand, cannot reach perfect conversion: it has an aperture for the entering solar radiation, and radiation emitted from the receiver's hot internal surfaces will be able to escape through this aperture. If we model the ideal receiver as a blackbody with aperture area A and uniform temperature T_{rec}, then the power lost due to this thermal emission is: $A\sigma(T_{rec}^4 - T_{amb}^4)$, where T_{amb} is the ambient temperature and σ is the Stefan-Boltzmann constant. The power entering the receiver aperture is: $ACI\eta_{opt}$, where I is the flux of direct sunlight. The ideal receiver efficiency is then

$$\eta_{rec}(C,T) = 1 - \frac{\sigma(T_{rec}^4 - T_{amb}^4)}{CI}. \tag{2}$$

The ideal power block includes the best possible heat engine, *i.e.*, an engine operating at Carnot efficiency; and an ideal generator with efficiency 1:

$$\eta_{PB}(C,T) = 1 - \frac{T_{amb}}{T_{rec}}. \tag{3}$$

Combining all three ideal subsystems, the overall plant efficiency is then:

$$\eta(C,T) = \left(1 - \frac{\sigma(T_{rec}^4 - T_{amb}^4)}{CI}\right)\left(1 - \frac{T_{amb}}{T_{rec}}\right). \tag{4}$$

The ideal plant efficiency is shown in fig. 1(b) as a function of receiver temperature for several levels of concentration [1]. $C = 50$ is representative of linear concentrators such as the parabolic trough and Linear Fresnel, $C = 500$ and $C = 1000$ are representative of solar towers, and $C = 3000$ can be achieved in a parabolic dish or a tower with secondary concentration. For each concentration, there is an optimal temperature, representing the best compromise between the thermal emission loss of the receiver and the thermodynamic loss of the heat engine. As the concentration increases, the maximum ideal efficiency increases, and the temperature required to achieve the optimum increases as well.

Real plants to not reach these ideal values, but they do follow the general trends shown in fig. 1(b). Therefore, technologies that enable increase in operating temperature and concentration have a good potential of increasing the plant conversion efficiency. It should be noted that such technologies may lead also to increase in plant installed cost, and therefore the overall impact of such advanced technologies on the levelized cost of energy should be investigated carefully.

2. – Optical performance

2˙1. *Mirrors.* – The concentrator component in a CSP plant is typically built of large mirrors mounted on a mechanism that tracks the direction of the sun. The optical losses include two different effects: reflection loss in the mirrors, which depend on the properties of the mirror material; and geometric losses that depend on the direction to the sun and on mirror and receiver geometry. The mirrors need to provide high specular reflectivity ($> 90\%$, *i.e.*, reflection loss less than 10%) and durability over the lifetime of a plant (> 20 years). To date, most of CSP mirrors have been produced from curved thick glass with a reflective back coating of silver, and protective layers behind the silver to prevent oxidation and mechanical damage in the reflective layer, as shown in fig. 2(a). These mirrors offer high specular reflectivity (typically $> 92\%$), and high durability (external glass surface is very robust even under external conditions). This

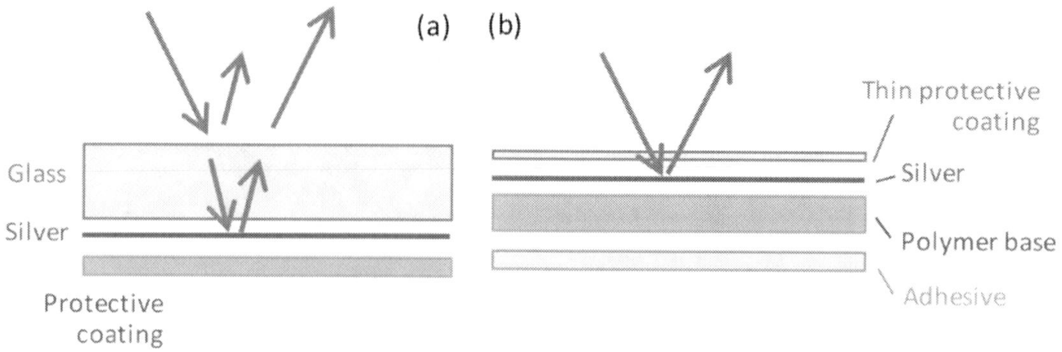

Fig. 2. – Schematic structure of mirrors: (a) back-surface silver on glass, (b) front surface on polymer.

solution requires curving of the glass at high accuracy, which can be done by thermal sagging (for short focal lengths) or elastic deformation (for long focal lengths), and can be expensive for both options. Thin glass (< 1 mm) with the same back silver coating is easier to curve and offers higher reflectivity ($> 95\%$), but such glasses are fragile and difficult to handle.

Several front-surface reflector systems have been proposed as alternatives to glass-based reflectors. One uses an enhanced aluminum or aluminum/silver coating on a polished aluminum substrate, with a top protective layer of anodization or polymer. However, these mirrors suffer from degradation and are not suitable for field deployment. Another solution is based on a polymer substrate coated with silver and protected by poly(methyl methacrylate) (PMMA), a transparent weather-resistant acrylic, and with a back adhesive for attaching to a substrate. The geometric shape of the mirror can be formed with a metal substrate, which is easier compared to bending glass, and also can provide structural rigidity after curving. These mirrors show promise, and some manufacturers claim to have achieved the needed high reflectivity ($> 93\%$) and durability (> 30 years in accelerated testing) to replace the reliable old solution of a glass back-surface reflector [2]. However, other evaluations are more skeptical [3], and most commercial CSP plants continue to use glass mirrors.

2·2. *Geometric losses.* – This category includes several effects related to the geometry of the concentrators. The geometric inaccuracy in the manufacturing and assembly of the optical surfaces may cause some of the reflected radiation to miss the receiver aperture, a loss often called “spillage”. Misalignment of the concentrator due to finite accuracy of the tracking mechanism can also cause spillage. Practical concentrators need to be very accurate, but the high cost of achieving very high accuracy leads to a compromise where some spillage is allowed.

Concentrators track to follow the direction towards the sun, but only the parabolic dish is able to fully accomplish this goal. Linear concentrators lack tracking in a second axis, and heliostats have to point away from the sun in order to send reflected radiation towards the receiver on the tower. In these cases, the collected radiation power is $IA\cos\theta$

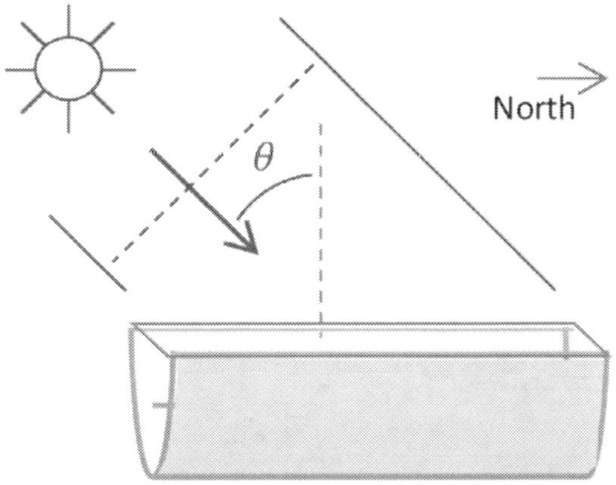

Fig. 3. – Schematic of the effect of incidence angle on the effective collection area of a parabolic trough.

where A is the collector aperture area, and θ is the incidence angle on the collector aperture (shown in fig. 3 for a typical geometry of a parabolic trough). As the incidence angle increases, the amount of collected power is decreased. This is often called the "cosine loss".

In a field of many concentrator units, mutual shading among neighboring units may cause a loss of incident radiation. In heliostat fields, the reflected radiation may be lost also on its way from the heliostat to the tower, and this is called a blocking loss. These losses vary in time due to the apparent motion of the sun. Shading and blocking losses can be reduced by increasing the distances among the concentrator units in the field, as can be seen in the heliostat field of fig. 4. This typically leads to a ratio of concentrator area to land area of up to 1:4, which means poor use of available land area.

Fig. 4. – PS10 heliostat field and tower, and part of PS20 field, near Seville, Spain (credit: Álvaro C.E., CC-BY-SA-3.0).

The design of concentrator fields and the optimization of the concentrators' geometry are usually performed with sophisticated optical simulation software [4]. Typical concentration values in practice are 30–80 for linear concentrators (trough, linear Fresnel) and 300–1,000 for point-focus devices (tower, parabolic dish). The optical efficiency for large scale concentrators is usually in the range of 60–75%, when averaged over annual operation (recall that some of the optical losses vary with time due to the apparent motion of the sun). An exception is the parabolic dish, which can achieve higher efficiency of > 85% and high concentration ratios even in large-scale industrial plants.

3. – Receiver performance

3·1. *Tubular receivers.* – The most common CSP receiver is a metal tube, illuminated by the concentrated solar radiation, with internal flow of a heat transfer fluid (HTF) to be heated. The energy balance of the receiver tube may be expressed in terms of incident *vs.* lost flux (power per unit external surface area of the tube). The net flux q_{net} available for heating the internal fluid is

$$q_{net} = q_{in} - q_{ref} - q_{em} - q_{conv} - q_{cond}, \tag{5}$$

$q_{in} = C \cdot I$ is the incident flux on the receiver. $q_{ref} = \rho_{rec}\, q_{in}$ is the flux reflected from the tube, where ρ_{rec} is the total reflectivity of the tube surface. $q_{em} = \epsilon_{rec}\, \sigma(T_{rec}^4 - T_{amb}^4)$ is the thermal emission loss, where ϵ_{rec} is the surface total emissivity. $q_{conv} = h(T_{rec} - T_{amb})$ is the convection loss from the exposed surface of the tube to the ambient air, where h is the effective convection coefficient. $q_{cond} = U(T_{rec} - T_{amb})$ is the conduction loss through any part of the tube surface that is insulated, where U is the heat transfer coefficient. The different loss mechanisms are shown schematically in fig. 5(a).

The receiver efficiency is then:

$$\eta_{rec} = \frac{q_{net}}{q_{in}} = 1 - \rho_{rec} - \frac{\epsilon_{rec}\, \sigma(T_{rec}^4 - T_{amb}^4) + (h + U)(T_{rec} - T_{amb})}{CI}. \tag{6}$$

Figure 5(b) shows part of a typical tubular receiver for line-focus concentrators, sometimes called the heat collection element (HCE). It includes an internal stainless steel pipe with a dark coating on the exterior of the pipe, an external glass pipe that maintains a vacuum around the steel pipe, a bellows at the end of the pipe to compensate for differences in thermal expansion between the glass and the metal, and a getter (not shown) to maintain the vacuum by absorbing gases that permeate into the evacuated space. These features contribute to reduce the various losses and increase the receiver efficiency, as discussed below.

Tubular receivers for point-focus concentrators have an arrangement of pipes carrying the HTF in a compact structure at the focal region of the concentrator. Typical structures include external receivers, where arrays of pipes are placed around a cylindrical volume on top of the tower, and cavity receivers, where arrays of tubes are located along the

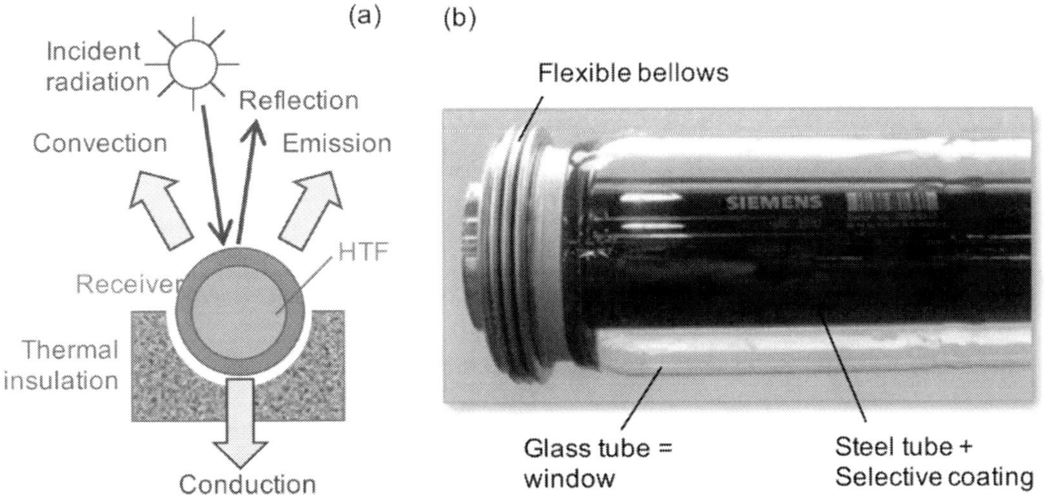

Fig. 5. – (a) Schematic showing the different losses from a tubular receiver; (b) a receiver for a parabolic trough collector (HCE, or vacuum tube).

inner walls of an insulated cavity. The tubes are usually made from a high-temperature-resistant steel or a nickel alloy such as Inconel. The metal is coated with a black coating or paint, but without the additional features of the linear focus receiver such as glass envelope and vacuum.

3˙2. *Reflection, emission and selective surfaces.* – The absorbing surface of the receiver should satisfy multiple requirements: it should absorb as much as possible of the incident solar radiation (low reflectivity); it should emit as little thermal radiation as possible (low emissivity); and it should also have a low convection loss. this combination of requirements is satisfied by two means applied to the receiver tube shown in fig. 5(b). The first is vacuum —an external transparent glass tube encloses the receiver tube, and the space between the tubes is evacuated. In the absence of air, the convection loss from the receiver tube is eliminated: $h = 0$ in eq. (6).

The second advanced element in the receiver is a selective coating [5]: it has high spectral emissivity for radiation in the solar spectrum (wavelengths of up to about $2\,\mu$m), and low spectral emissivity for wavelengths that characterize the emission from the hot receiver (2–10 μm). Figure 6(a) shows the principle of an ideal spectrally selective surface. Due to the difference between the solar spectrum and the spectrum of emitted thermal radiation, the integrated total absorptivity and total emissivity can be very different. A wide range of materials and microstructures can produce such spectrally selective behavior [6]. The most popular solutions are cermets (composites of ceramic and metal) with several thin layers, or a layer of ceramic with metal particles, over a back metallic IR reflector, as shown in fig. 6(b). In a real selective coating, the change in the spectral emissivity is gradual and not a step change, as shown in fig. 7(a) for a typical Cr/Cr_2O_3 coating [7]. This however is sufficient to create the spectral selectivity effect.

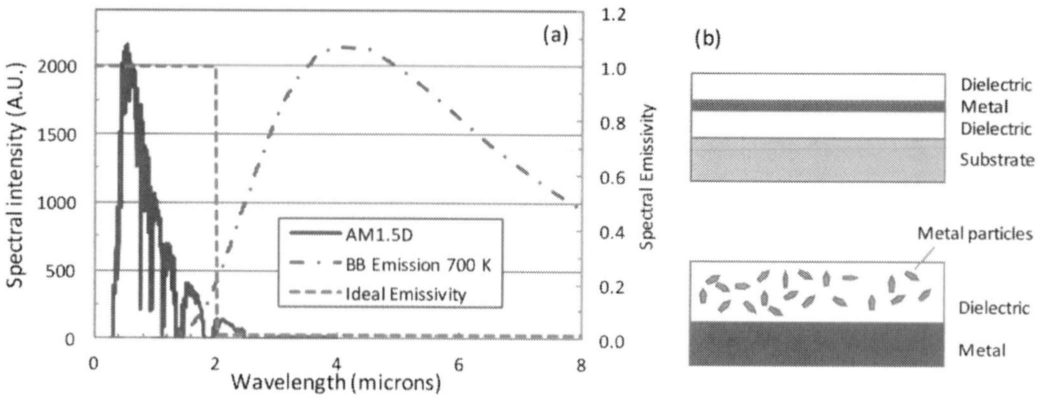

Fig. 6. – (a) Spectra of sunlight (AM1.5D) and of blackbody emission at 700 K, with the spectral emissivity of an ideally selective absorber surface. (b) Typical structures of Cermet selective coatings.

A typical commercial selective coating provides a high total absorptivity of 0.93–0.96 for solar radiation, together with a low overall emissivity of about 0.05 at room temperature and 0.1–0.15 at the receiver's operating temperature of 400–500 °C. Figure 7(b) shows the resulting efficiency of a typical parabolic trough receiver: from the measurement of the overall efficiency from sunlight to heat, and a separate measurement of the optical efficiency, the receiver efficiency can be identified and it is 95% at receiver temperature of 350 °C, declining to 83% at 500 °C.

The main limitation of the selective coatings is stability at high temperature, particularly if the coating is exposed to air, and they tend to fail if the vacuum in the HCE is breached. Efforts are ongoing to develop high-temperature air stable coatings, however the solutions found so far contain relatively expensive materials and are not yet practical for commercial applications [6].

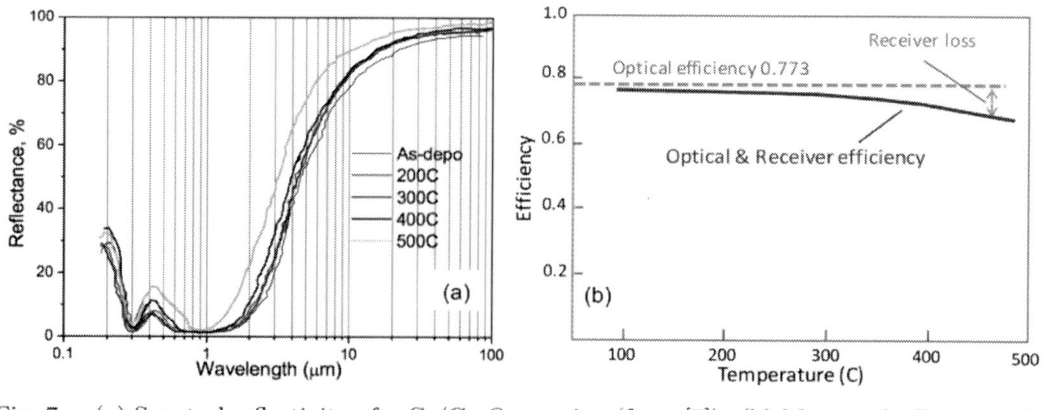

Fig. 7. – (a) Spectral reflectivity of a Cr/Cr_2O_3 coating (from [7]). (b) Measured efficiency of a parabolic trough collector *vs.* temperature (adapted from [8]), showing the contributions of the optical and receiver losses.

DOI: 10.1051/epjconf/201714800009

TABLE I. – *Typical power block efficiencies in steam plants as a function of operating temperature and pressure.*

Temperature (C)	Pressure (bar)	Efficiency
400	100	32%
530	170	37%
590	250	45%

4. – Heat engine efficiency

The two most common power conversion cycles in conventional power plants are the steam-based Rankine cycle and the air-based open Brayton cycle (also called the gas turbine cycle). In both cycles, mechanical power is produced by manipulation of pressure and temperature of the working fluid: water or air. A detailed description of the operation of these standard cycles can be found in many thermodynamics textbooks. A combination of these two cycles, the combined cycle (CC), uses the hot exhaust stream from the gas turbine to generate the steam for the Rankine cycle, thus exploiting a larger fraction of the heat input and reaching a higher efficiency. Currently, only the steam cycle is used in CSP plants, although much effort has been invested in developing solar technology for gas turbines as well.

The conversion efficiency of the heat engine (or the power block, containing also the generator attached to the heat engine) depends on the operating temperature. Higher temperature leads in general to higher efficiency, consistent with the ideal limit of Carnot efficiency discussed above. The selection of the heat transfer fluid poses limitations on achievable temperature. In most parabolic trough plants today the HTF is a synthetic oil such as Therminol VP-1 that can operate up to 400 °C. The hot HTF is used to generate steam for the power generation cycle in a boiler, and the steam temperature is then limited to less than 400 °C. This leads to relatively low conversion efficiency, as shown in table I. In some tower plants the HTF is a molten salt (mixture of $NaNO_3$ and KNO_3) that allows a higher temperature of about 550 °C, leading to a higher power block conversion efficiency. The schematic of a CSP plant with molten salt as HTF is shown in fig. 8. Molten nitrate salt is a popular choice for current CSP plants, but it freezes at about 240 °C, leading to major difficulties in salt management and freeze protection during system outages. Formulating alternate salts, or other liquids, with a lower freezing temperature is then a major challenge for materials research, considering that the list of requirements also includes stability at high temperature (at least $>$ 500 °C), reasonably high thermal conductivity and specific heat, reasonably low viscosity across the temperature range, chemical compatibility with standard tube materials, and obviously very low cost. Other proposals for alternate HTFs include pressurized CO_2 and some ionic liquids, but these are still in the early research phase.

The plant layout can be simplified if the HTF circuit is eliminated, and the entire process is done with water and steam. However, this requires the receiver tubes to accept steam at very high pressures, as shown in table I. This is not desirable since the high

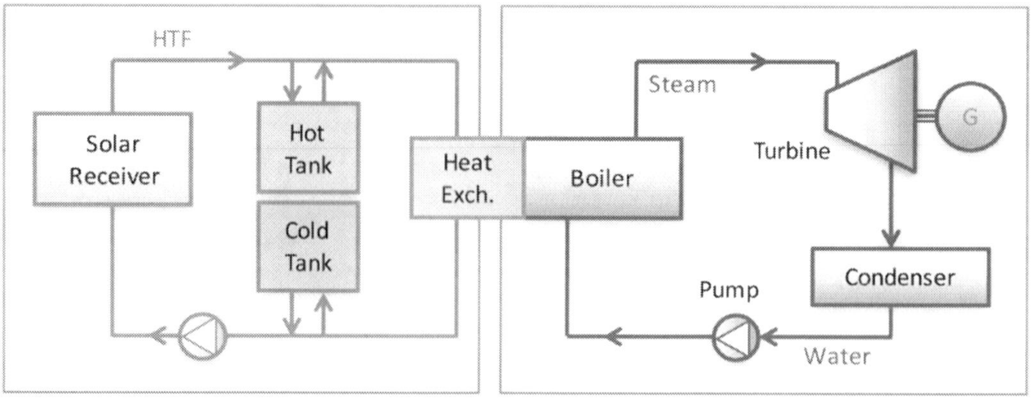

Fig. 8. – Schematic of a CSP plant with indirect heating, containing a solar circuit with a HTF (*e.g.* molten salt) and storage of hot fluid, and a steam circuit that operates the heat engine.

temperature reduces the mechanical strength of the tubes and requires the design of very expensive tubes. In addition, the operation and control of a large array of steam tubes under varying solar input conditions, in particular in a parabolic trough field, is quite difficult. There are several tower plants that operate with direct steam, but this approach is less favored and molten salt is preferred, in particular due to the inherent ability to include storage in the circuit. Direct steam has also been attempted in parabolic troughs [9], but it is currently used only in Linear Fresnel plants.

5. – Overall CSP plant efficiency

Table II shows the main features for two common examples of current CSP plants. One represents the parabolic troughs with low concentration and synthetic oil HTF, which are the most common plant configuration, existing since the 1980s. The second represents solar towers with higher concentration and molten salt as HTF, which have been constructed during the last decade. For both types of plants, the annual average efficiency of conversion from sunlight to electricity is less than 20%. This efficiency is similar to photovoltaic systems, but the complexity and cost of the CSP plants are much higher, and therefore the thermal conversion technologies are not able to reach the commercial success enjoyed by PV systems.

Table II. – *Main features and plant efficiencies for two common configurations of current CSP plants.*

Concentration	Low	High
HTF: Heat transfer fluid	Synthetic oil Biphenyl/diphenyl oxide	Molten nitrate salt $NaNO_3/KNO_3$
Maximum temperature	400 °C	550 °C
Cycle efficiency	0.32	0.38
Plant efficiency	0.14	0.18

DOI: 10.1051/epjconf/201714800009

Achieving a more competitive CSP technology requires progress in two paths. One is cost reduction in every step of the manufacturing, installation and operation of the plants. This path can produce a series of gradual small improvements that by themselves are not sufficient to reach sufficient competitiveness [10]. The second path is development of higher performance solutions that will offer much higher conversion efficiency, such as new thermodynamic cycles, new HTF materials, etc. However this path is long and difficult, and requires not only a concept that works, but also an industrial development process that will demonstrate feasibility under field conditions, reliability over decades of operation, and low cost of the proposed new concept.

6. – Challenges for CSP

CSP plants are already operating on a large scale in many locations around the world. However, currently CSP offers moderate conversion efficiencies and uncompetitive costs, and therefore requires significant government incentives. Many opportunities are available for improving CSP: increasing the operating temperature, which, in turn, would increase the conversion efficiency and reduce specific costs (per unit energy); reducing component and plant costs by developing improved materials and manufacturing practices; and adding thermal storage to provide true dispatchability. These improvements require major advances in materials, thermodynamics, heat transfer, and other disciplines, as well as further development and demonstration. It is expected that improved CSP technologies will reach grid parity around 2025, leading to a better competitive position in the market, and to implementation on a much larger scale.

REFERENCES

[1] Fletcher E. A. and Moen R. L., *Science*, **197** (1977) 1050, doi:10.1126/science.197.4308.1050.
[2] DiGrazia M., Jorgensen G., Gee R. and Bingham C., *Service Life Prediction for ReflecTech®PLUS Mirror Film, World Renewable Energy Forum, Denver, May 2012.*
[3] Kennedy C. E. and Terwilliger K., *J. Sol. Energy Eng.*, **127** (2005) 262.
[4] Li L., Coventry J., Bader R., Pye J. and Lipiński W., *Opt. Express*, **24** (2016) A985, doi:10.1364/OE.24.00A985.
[5] Tabor H., *Proc. Natl. Acad. Sci. U.S.A.*, **47** (1961) 1271, doi:10.1073/pnas.47.8.1271.
[6] Kennedy C. E., *Review of Mid- to High- Temperature Solar Selective Absorber Materials Review of Mid- to High- Temperature Solar Selective Absorber Materials*, Technical Report (National Renewable Energy Lab., Golden, CO) 2002, `https://ases.conference-services.net/resources/252/2859/pdf/SOLAR2012_0698_full%20paper.pdf`.
[7] Cao F., McEnaney K., Chen G. and Ren Z., *Energy Environ. Sci.*, **7** (2016) 1615, doi:10.1039/c3ee43825b.
[8] SkyTrough® Thermal Efficiency (2010).
[9] Zarza E., Valenzuela L., Leon J., Hennecke K., Eck M., Weyers H. D. *et al.*, *Energy*, **29** (2004) 635.
[10] Pitz-Paal R., Amin A., Oliver Bettzuge M., Eames P., Flamant G., Fabrizi F. *et al.*, *J. Sol. Energy Eng.*, **134** (2012) 24501, doi:10.1115/1.4006390.

Thermoelectricity for future sustainable energy technologies

ANKE WEIDENKAFF(*)

Lehrstuhl für Chemische Materialsynthese, Institut für Materialwissenschaft
Universität Stuttgart - Heisenbergstraße 3, 70569 Stuttgart, Germany

Summary. — Thermoelectricity is a general term for a number of effects describing the direct interconversion of heat and electricity. Thermoelectric devices are therefore promising, environmental-friendly alternatives to conventional power generators or cooling units. Since the mid-90s, research on thermoelectric properties and their applications has steadily increased. In the course of years, the development of high-temperature resistant TE materials and devices has emerged as one of the main areas of interest focusing both on basic research and practical applications. A wide range of innovative and cost-efficient material classes has been studied and their properties improved. This has also led to advances in synthesis and metrology. The paper starts out with thermoelectric history, basic effects underlying thermoelectric conversion and selected examples of application. The main part focuses on thermoelectric materials including an outline of the design rules, a review on the most common materials and the feasibility of improved future high-temperature thermoelectric converters.

1. – Introduction

Increasing efficiency of any kind of technical process is imperative from both an environmental and an economic standpoint. Since a large proportion (60–70%) of the worldwide-applied energy carriers is converted into unused heat, recovery of at least part of this "waste" heat could lead to extensive savings. Thermoelectric (TE) converters,

(*) E-mail: weidenkaff@imw.uni-stuttgart.de

DOI: 10.1051/epjconf/201714800010

which directly convert thermal energy into electric energy (Seebeck effect) appear particularly suitable for this purpose. They are all the more attractive due to their simplicity, silence and reliability due to the absence of moving parts. They also enable the direct and efficient conversion of abundant solar and geothermal heat. However, growth of thermoelectric conversion requires the development of effective, stable, sustainable and cost-efficient materials [1].

The collective term "thermoelectricity" summarizes various effects discovered almost 200 years ago. In 1822 the German-Baltic physicist and physician Thomas Johann Seebeck discovered an effect which was named after him [2]. Seebeck failed to recognize the actual cause of his findings and it was the Danish physicist and chemist Hans Christian Ørsted who called it a "thermoelectric effect" for the first time [3]. The Seebeck effect consists in a voltage gradient as the result of a thermal gradient across a sample. This is the consequence of a higher absolute value of the average velocity of charge carriers in the warm end of the sample, which leads to an effective diffusion towards the cold end. In a steady state this can be expressed by

$$-\nabla U = S\nabla T, \tag{1}$$

where U is the voltage, S the Seebeck coefficient and T the absolute temperature.

Twelve years later the French watchmaker, Jean Charles Athanase Peltier, observed the reverse process [4]. But just as Seebeck Peltier misinterpreted his results and it was not until 1838 that another German-Baltic physicist Heinrich Friedrich Emil Lenz realized the true nature of the Peltier effect and proved it to be an autonomous phenomenon [5]. It is defined as

$$|\dot{Q}| = \Pi_{AB} \cdot I = (\Pi_A - \Pi_B) \cdot I, \tag{2}$$

where Q is the heat, Π are the Peltier coefficients of the materials A and B and I the electric current. Basically, the conditions of the Peltier effect are just the opposite of the Seebeck effect: in this case, an electric current induces heat transport, *i.e.* heat can be absorbed or released by applying a voltage.

Systematic studies on thermoelectricity were not further pursued until around 1885. Between 1909 and 1911 the German physicist and mathematician Edmund Altenkirch succeeded in calculating the potential efficiency of thermoelectric generators and the performance of Peltier coolers [6, 7] (fig. 1). Using his constant property model he already back then identified the parameters that make a good thermoelectric material: large Seebeck coefficient, high electrical conductivity (to minimize Joule heating due to electrical resistance) and low thermal conductivity (to minimize heat loss). In 1949 Ioffe developed the concept of the Figure of Merit, Z, describing the thermoelectric quality of a material which he presented in 1957 [8]:

$$Z = \frac{S^2\sigma}{\kappa}, \tag{3}$$

 DOI: 10.1051/epjconf/201714800010

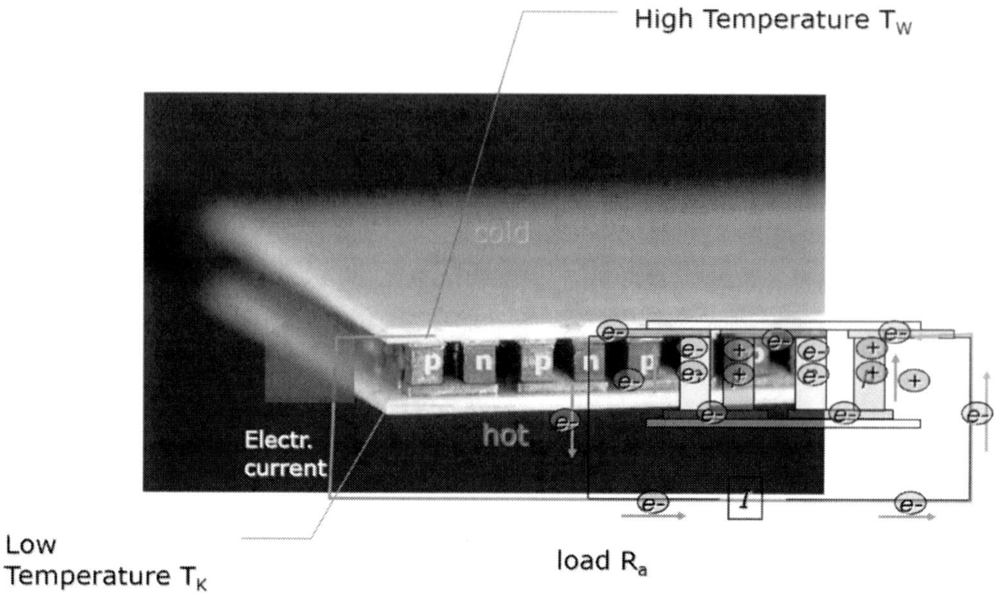

Fig. 1. – Thermoelectric generator (TEG) and Peltier cooler. The thermoelectric module is composed of an n-type and a p-type semiconducting material connected electrically in series through metallic electrical contact pads and thermally in parallel between ceramic ends.

where $S^2\sigma$ represents the power factor P_f, with S the Seebeck coefficient, σ the electrical conductivity and κ the thermal conductivity.

Using the dimensionless product ZT,

$$ZT = \frac{S^2\sigma}{\kappa}T, \tag{4}$$

the efficiency of a thermoelectric generator (TEG), *i.e.* the ratio between the energy delivered at the load and the heat energy absorbed at the hot side, can be expressed in terms of the Carnot-efficiency η_{Carnot} (the upper efficiency limit of any heat engine) by

$$\eta_{TE} = \frac{(1+ZT)^{1/2}-1}{(1+ZT)^{1/2}+\frac{T_{cold}}{T_{hot}}} \cdot \eta_{Carnot}. \tag{5}$$

Figure 2 illustrates the relation between the maximum conversion efficiency and the ZT.

As mentioned above, eqs. (3) and (4) prove that suitable materials are characterized by a high Seebeck coefficient $|S|$, good electric conductivity σ and low thermal conductivity κ. However, the selective modification of just one parameter is complicated by their mutual dependence. An increased electric conductivity usually comes along with an enhanced thermal conductivity; a higher carrier concentration improves the electric conduction at the expense of a reduced Seebeck coefficient. Thus, the main objective is

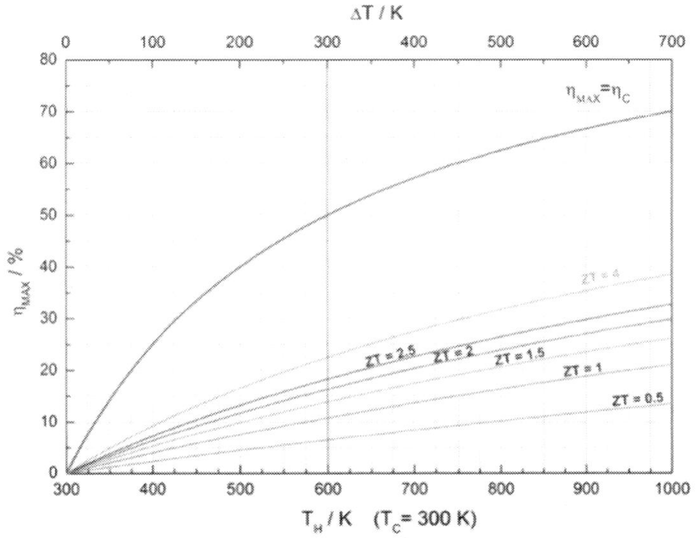

Fig. 2. – Relation between maximum conversion efficiency η_{MAX} and the thermoelectric figure of merit Z.

to develop materials with a charge carrier concentration, mass and mobility providing a high power factor P_f while at the same time suppressing the thermal conductivity (fig. 3). Moreover, it follows from eq. (4) that the economic fields of application arise in high-temperature environments.

The emergence of innovative synthesis and analysis methods together with the need of energy conservation resulted in a revival of thermoelectric research activities. Ioffe was one of the first to acknowledge the importance of alloying to reduce the thermal conductivity. Also, new nanostructuring techniques for the production of novel material classes with improved thermoelectric properties were developed. Since the mid-90s thermoelectric research has significantly increased. The focus is especially on high-temperature thermoelectrics and environmental-friendly and inexpensive materials.

2. – Examples of applications

Thermoelectric converters can be used for thermometry, power generation, heating and cooling purposes. The application generally requires p- and n-type thermoelectrics. For thermometry, the resistance at the contact is coordinated with the temperature by calibrating against a reference thermometer. For power generation or heating / cooling devices, the thermoelements are connected electrically in series and sandwiched between a hot and a cold heat exchanging system. If a temperature gradient is applied the charge carriers travel along the T gradient creating a current which can be used by a consumer. If a voltage is applied the same device can be used for heating and cooling, because heat is transported by the moving charge carriers.

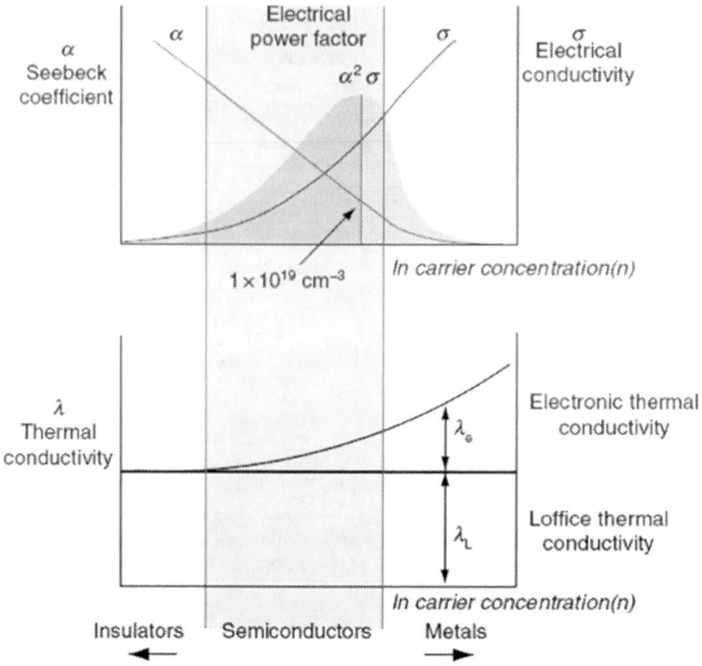

Fig. 3. – Schematic dependence of electrical conductivity, Seebeck coefficient, power factor and thermal conductivity on concentration of free carriers [9].

Wherever temperature gradients arise, thermoelectric devices can be applied. Owing to their simple design free of moving parts they offer durable and maintenance-friendly current supply. This can be of particular importance for unmanned remote facilities such as light and radio buoys, lighthouses, sensing stations for radar systems and pipeline monitoring equipment. An additional vast area of application is aerospace. Since the early 1960s, radioisotope thermoelectric generators (RTG) have been used as power supply for satellites and space crafts due to their longevity, reliability and small dimensions. They are the means of choice for missions to long for batteries or unfit for photovoltaics. Prominent examples include space probes of Pioneer, Voyager and Galileo missions and the power supply for scientific experiments by Apollo crews.

At present, application of this technique is largely confined to the above-mentioned niche products, since affordable and efficient TE materials are still lacking. The most significant improvements in performance are going to be expected at high temperature differences. Therefore, the key to a broader market would be inexpensive TE materials combining high conversion efficiency and adequate temperature stability.

A number of research projects have already demonstrated efficiency enhancements of technical installations by TEGs. In the project "Customized Hybrid Powertrains" (Cohyb), supported by Swisselectric (organisation of the Swiss electricity grid operators) together with the Swiss "Competence Center Energy and Mobility (CCEM)", TE recu-

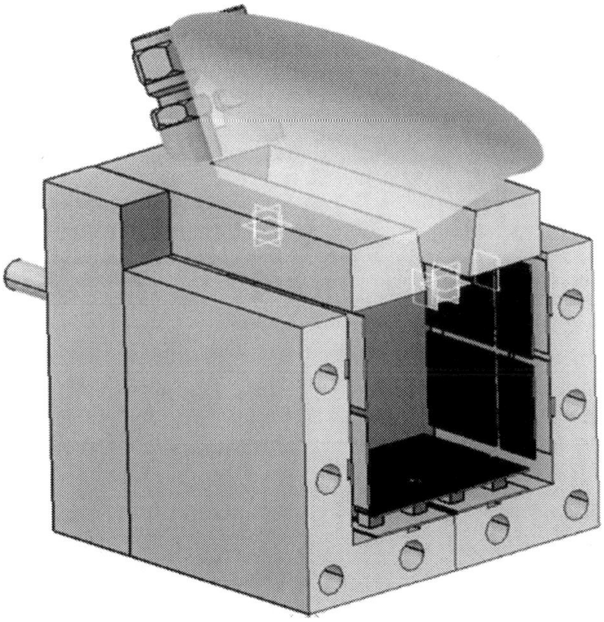

Fig. 4. – Solar thermoelectric cavity converter with 18 thermoelectric oxide modules (TOM) and K-type thermocouples for temperature monitoring.

peration of $\sim$ 120 W from engine exhaust heat, corresponding to 3% fuel saving, was achieved [10]. The goal of the project "High Temperature Thermoelectric Converter for Electricity Generation in a SOFC System" (HITTEC), cofounded by CCEM and the Swiss Federal Office of Energy (SFOE), was to develop a novel high-temperature (HT - up to 1200 K) thermoelectric converter (TEC) to be integrated, for instance, in a solid oxide fuel cell system (SOFC). The all-oxide module showed excellent HT-stability and a power output of 50 W at temperature differences of ΔT = 400–500 K [11]. All-oxide modules where subject of research in another promising HT application. Thermoelectric conversion of simulated concentrated solar radiation was studied by means of a solar cavity converter (fig. 4).

The converter effectively captured the solar heat power and reduced re-radiation to 4%. In this way the solar-to-electricity conversion efficiency could be increased by 62% compared to a module subjected to direct solar irradiation [12-14].

Subject to the development of good cost-benefit materials many more —particularly high-temperature— applications are conceivable. These include the conversion heat of geothermal or automotive origin or from machines of energy-intensive industries (steel, paper, glass, cement).

3. – Materials

3'1. *Design rules.* – Any effort to develop high-performance TE materials has to aim at an increasing ZT, *i.e.* a power factor $P_f(S^2\sigma)$ as high and a thermal conductivity κ

as low as possible (cf. eq. (4)). Beneficial effects on these parameters can be achieved by the following measures:

- Low thermal conductivity is a consequence of increased phonon scattering: This is achieved by mass fluctuation, heavy atoms, complex crystal structures or enhanced grain boundary and Umklapp scattering.
- A high Seebeck coefficient is associated with a steep rise of the density of states (DoS) at the Fermi level. This can be realized by tuning the band structure and band gap with low-dimensional systems, correlated electrons, low minority carriers or carriers with high effective mass.
- A good electrical conductivity relies on a high charge carrier concentration and mobility. This requires narrow-band-gap semiconductors and low electron scattering. Good electronic conductivity can be found in materials with strongly correlated electrons.

In addition, good temperature stability in air would be advantageous as the conversion efficiency is also a function of the temperature gradient.

Unfortunately, the complex physical relationships between these quantities illustrated by eqs. (6) to (8) lead to conflicting targets:

$$S = \frac{8\pi^2 k_B^2}{3eh^2} m^* T \left(\frac{\pi}{3n}\right)^{2/3}, \tag{6}$$

$$\sigma = ne\mu = \frac{ne^2\tau}{m^*}, \tag{7}$$

$$\kappa_{tot} = \kappa_{ph} + \kappa_{el} = \kappa_{ph} + L\sigma T, \tag{8}$$

where k_B is the Boltzmann constant, m^* the density of states effective mass, h the Planck constant, n the carrier concentration, e the elementary charge, μ the carrier mobility, τ the relaxation time, κ_{tot} the total thermal conductivity, κ_{ph} the phonon (or lattice) thermal conductivity, κ_{el} the electronical thermal conductivity, and L the Lorenz number.

Equation (8) shows that lowering the phonon part κ_{ph}, whose influence usually outweighs that of the electronic part κ_{el}, to values exhibited by glass-like structures would be the most potent way to reduce the overall thermal conductivity κ_{tot}. For the phonon thermal conductivity κ_{ph} one has

$$\kappa_{ph} \propto C v_s l_{ph}, \tag{9}$$

where C is the heat capacity per volume, v_s the phonon velocity (speed of sound), and l_{ph} the phonon mean free path. Therefore possible approaches towards increased phonon scattering are the generation of grain boundaries or defects ($\downarrow l_{ph}$), complex crystal or "misfit" structures and using high-atomic-weight elements ($\downarrow v_s$).

On the other hand, charge carrier mobility, *i.e.* the electric conductivity σ, increases with the crystallinity of the material. Thus, the challenge is to effectively scatter the phonons while preserving the electrical concuctivity. The compromise solution is the concept of a "phonon-glass electron-crystal" serving as a guideline for the design of thermoelectrics with improved properties [15].

What is more, the materials shall consist of inexpensive, abundant, and non-toxic elements.

3·2. *Materials review.* – Chalcogenides such as tellurides (*e.g.* $AgPb_{18}SbTe_{20}$(LAST)) are the currently best and most broadly applied thermoelectric materials. They operate in the low-temperature range up to maximum temperatures of ca. 500 K with reproducible ZT values in the range of $0.7 < ZT < 1.5$. With a suitable nano-substructure, *i.e.* a 2D crystal structure, high power factors are common in *e.g.* Bi_2Te_3 [16] and Bi_2Te_3/Sb_2Te_3 superlattices [17]. Recently, sulfide-based minerals such as tetrahydrites (CuSb sulfides) have been considered alternative, low-cost candidates for thermoelectric power generation [18]. TiS_2 nanosheets (sulphene) offer the possibility of a two-dimensional confinement of electrons and phonons in favor of very high ZT values [19].

Cage structures such as skutterudites or filled skutterudites (*e.g.* $CeFe_4Sb_{12}$) and Si- or Ge-based clathrates reveal a complex crystal structure with "rattling" atoms and very low thermal conductivity [20, 21]. They can be nanostructured by *e.g.* melt spinning or nano precipitation methods.

(Half)Heusler compounds are narrow-gap semiconductors (*e.g.* TiNiSn) with high effective mass charge carriers leading to large Seebeck coefficients of up to $S = 400\,\mu VK^{-1}$ at room temperature. They are produced by co-melting the metals in an arc or induction furnace and a subsequent homogenation procedure. Ta substitution in these Heusler-based thermoelectrics suppresses the minority carriers (impurity band) and enhances the density-of-states effective mass $m_C{}^*$ and the charge carrier mobility μ in $Zr_{0.3}Hf_{0.7-x}Ta_xNiSn$ [22, 23]. Nanostructuration can be achieved by spinodal decompositions, ball milling or nanoinclusions of full Heusler precipitates.

Oxides are stable in oxidizing environments (air) at high temperatures. Since the discovery of high Seebeck coefficient combined with low electrical resistivity of $NaCo_2O_4$ single crystals in the 1990s [24] at the latest, oxides have been recognized as a very important alternative material class for energy conversion in high-temperature environments. $NaCo_2O_4$ is a p-type cobalt oxide with a layered hexagonal structure. Materials with layered structures come close to a "phonon-glass-electron-crystal", the ideal configuration for most effective thermoelectric materials. The lower 2D-arrangement of alternating layers is conducive both to phonon scattering and to charge carrier conduction and emphasizes the significance of the dimensionality for the development of innovative thermoelectric materials.

In addition, particular attention has been directed to transition metals-containing perovskite-type oxides owing their attractive properties and their non-poisonous and abundant elements [25-31]. Many ternary compounds with the general formula ABX_3 adopt the perovskite structure, which is ideally made up of BX_6 octahedra (X=O,N) joined at corners with the A cation located in the center of eight octahedra (fig. 5).

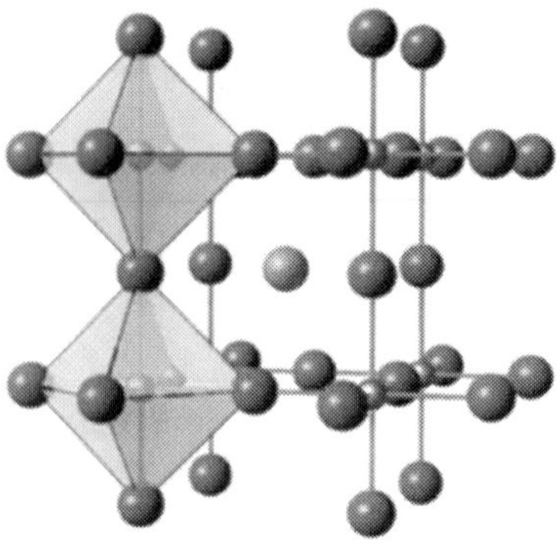

Fig. 5. – Ideal perovskite structure with each B cation (grey) surrounded by six anions (violet) forming an BX_6 (X=O,N) coordination octahedron. The A cation (blue) is located in the center of eight octahedra.

The perovskite structure is very flexible in terms of chemical variations allowing adjustment of the band structure (DoS / band gap management) and physical properties (electronic correlation, spin entropy) by means of the chemical composition. Defects and local non-stoichiometry in the crystal structure are common and frequently tolerated leading to stable distorted structures, super structures, layered structures, polar structures, defects or non-stoichiometric (A-cation deficient, B-cation deficient, anion deficient) regenerative compounds. Perovskites show a wide variety of interesting properties, because many different compositions and also mixed-valence compounds can be stabilized in this flexible crystal structure. They have a particularly huge potential for many energy conversion processes due to their insulating, p- or n-type semiconducting, metallic or even superconducting transport properties. Complex oxides with perovskite-type structures offer a suitable, correlated electrons system for improving the thermopower by spin-entropy contribution. Well-known and useful perovskites are cobaltates, titanates, manganates, niobates and tantalates.

Silicides such as, for example, $MnSi_{1.75}$ and solid solutions of compounds based on Mg_2X (X = Si, Ge, and Sn) are environmentally friendly materials with high ZT values and low weight [32].

Organic and hybrid materials are interesting for room temperature applications with a large surface area of the heat exchanger system. They are potentially low-cost and flexible materials. Conducting polymers including polyaniline, polypyrrol and poly(3,4-ethylenedioxythiophene) (PEDOT) exhibit attractive thermoelectric properties and will be studied for thermoelectric applications [33].

New thermoelectric materials for energy conversion will provide an enhancement of the energy density, better thermal stability and efficiency of heat to electricity converters. The design of new materials is based on band gap, size and defect control by innovative synthesis processes. Furthermore, research on the large thermopower in correlated electronic systems and the low thermal conductivity due to grain boundaries hindering phonon transport has to be continued. Scalable, cost-efficient synthesis methods for environmentally friendly nanostructured materials and segmented converters using materials combinations will enable thermoelectric conversion at $T > 700\,^{\circ}\mathrm{C}$ improving the power output.

REFERENCES

[1] Rowe D. M., "Thermoelectric Waste Heat Recovery as a renewable energy source", *Int. J. Innov. Energy Syst. Power*, **1** (2006) 13.
[2] Seebeck T., "Magnetische Polarisation der Metalle und Erze durch Temperatur-Differenz", *Abh. K. Preuss. Akad. Wiss. Berlin* (1822) 265.
[3] Ørsted H. C., "Nouvelles experiences de M. Seebeck sur les actions èlectro-magnetiques", *Ann. Chim. Phys.*, **22** (1823) 375.
[4] Peltier J. C. A., "Nouvelles expériences sur la caloricité des courans électriques", *Ann. Chim. Phys.*, **2** (1834) 371.
[5] Lenz E., "Einige Versuche im Gebiete des Galvanismus", *Ann. Phys. (Leipzig)*, **120** (1838) 342, DOI: 10.1002/andp.18381200612.
[6] Altenkirch E., "Über den Nutzeffekt der Thermosäule", *Phys. Z.*, **10** (1909) 560.
[7] Altenkirch E., "Elektrothermische Kälteerzeugung und reversible elektrische Heizung", *Phys. Z.*, **12** (1911) 920.
[8] Ioffe A. F., *Semiconductor Thermoelements and Thermoelectric Cooling* (Infosearch, Ltd., London) 1957.
[9] Rowe D. M. (Editor), *Thermoelectrics Handbook. Macro to Nano* (CRC/Taylor & Francis, Boca Raton, 2006).
[10] Guzzella L., "Cohyb - Customized Hybrid Powertrains", in *CCEM Annual Activity Report 2013* (Competence Center Energy and Mobility (ccem.ch), Paul Scherrer Institut, Villigen (CH)) 2014, pp. 26–29.
[11] Heel A. and Populoh S., "HITTEC - High Temperature Thermoelectric Converter for Electricity Generation in a SOFC System", in *CCEM Annual Activity Report 2015* (Competence Center Energy and Mobility (ccem.ch), Paul Scherrer Institut, Villigen (CH)) 2016, pp. 17–19.
[12] Tomes P., Trottmann M., Suter C., Aguirre M. H., Steinfeld A., Haueter P. and Weidenkaff A., "Thermoelectric Oxide Modules (TOMs) for the Direct Conversion of Simulated Solar Radiation into Electrical Energy", *Materials*, **3** (2010) 2801, DOI: 10.3390/ma3042801.
[13] Tomes P., Suter C., Trottmann M., Steinfeld A. and Weidenkaff A., "Thermoelectric oxide modules tested in a solar cavity-receiver", *J. Mater. Res.*, **26** (2011) 1975, DOI: 10.1557/jmr.2011.125.
[14] Suter C., Tomeš P., Weidenkaff A. and Steinfeld A., "A solar cavity-receiver packed with an array of thermoelectric converter modules", *Sol. Energy*, **85** (2011) 1511, DOI: 10.1016/j.solener.2011.04.008.
[15] Slack G., "New Materials and Performance Limits for Thermoelectric Cooling", in *CRC Handbook of Thermoelectrics/CRC handbook of thermoelectrics*, edited by Rowe D. M. (CRC Press, Boca Raton, FL) 1995, pp. 407–440.
[16] Xie W., He J., Kang H. J., Tang X., Zhu S., Laver M., Wang S., Copley J. R. D., Brown C. M., Zhang Q. and Tritt T. M., "Identifying the specific nanostructures responsible for the high thermoelectric performance of (Bi,Sb)2Te3 nanocomposites", *Nano Lett.*, **10** (2010) 3283.
[17] Venkatasubramanian R., Siivola E., Colpitts T. and O'Quinn B., "Thin-film thermoelectric devices with high room-temperature figures of merit", *Nature*, **413** (2001) 597, DOI: 10.1038/35098012.

[18] Lu X. and Morelli D. T., "Rapid synthesis of high-performance thermoelectric materials directly from natural mineral tetrahedrite", *Magn. Res. Chem.*, **3** (2013) 129, DOI: 10.1557/mrc.2013.26.

[19] Zhang R.-z., Wan C.-l., Wang Y.-f. and Koumoto K., "Titanium sulphene: two-dimensional confinement of electrons and phonons giving rise to improved thermoelectric performance", *Phys. Chem. Chem. Phys.*, **14** (2012) 15641, DOI: 10.1039/c2cp42949g.

[20] Tritt T. M. and Subramanian M. A., "Thermoelectric Materials, Phenomena, and Applications: A Bird's Eye View", *MRS Bull.*, **31** (2006) 188, DOI: 10.1557/mrs2006.44.

[21] Eilertsen J., Surace Y., Balog S., Sagarna L., Rogl G., Horky J., Trottmann M., Rogl P., Subramanian M. A. and Weidenkaff A., "From Occupied Voids to Nanoprecipitates: Synthesis of Skutterudite Nanocomposites in situ", *Z. Anorg. Allg. Chem.*, **641** (2015) 1495, DOI: 10.1002/zaac.201500137.

[22] Gałązka K., Populoh S., Xie W., Yoon S., Saucke G., Hulliger J. and Weidenkaff A., "Improved thermoelectric performance of (Zr0.3Hf0.7) NiSn half-Heusler compounds by Ta substitution", *J. Appl. Phys.*, **115** (2014) 183704.

[23] Zeier W. G., Schmitt J., Hautier G., Aydemir U., Gibbs Z. M., Felser C. and Snyder G. J., "Engineering half-Heusler thermoelectric materials using Zintl chemistry", *Nat. Rev. Mater.*, **1** (2016) 16032, DOI: 10.1038/natrevmats.2016.32.

[24] Terasaki I., Sasago Y. and Uchinokura K., "Large thermoelectric power in NaCo2O4 single crystals", *Phys. Rev. B*, **56** (1997) R12685, DOI: 10.1103/PhysRevB.56.R12685.

[25] Terasaki I., Iwakawa M., Nakano T., Tsukuda A. and Kobayashi W., "Novel thermoelectric properties of complex transition-metal oxides", *Dalton Trans.*, **39** (2010) 1005, DOI: 10.1039/B914661j.

[26] Hashimoto H., Kusunose T. and Sekino T., "Influence of ionic sizes of rare earths on thermoelectric properties of perovskite-type rare earth cobalt oxides RCoO3 (R=Pr, Nd, Tb, Dy)", *J. Alloys Compd.*, **484** (2009) 246.

[27] Weidenkaff A., Robert R., Aguirre M. H., Bocher L., Lippert T. and Canulescu S., "Development of thermoelectric oxides for renewable energy conversion technologies", *Renew. Energy*, **33** (2008) 342, DOI: 10.1016/j.renene.2007.05.032.

[28] Weidenkaff A., Robert R., Aguirre M. H., Bocher L. and Schlapbach L., "Nanostructured thermoelectric oxides with low thermal conductivity", *Phys. Status Solidi Rapid Res. Lett.*, **1** (2007) 247, DOI: 10.1002/pssr.200701185.

[29] Koumoto K., Terasaki I. and Funahashi R., "Complex Oxide Materials for Potential Thermoelectric Applications", *MRS Bull.*, **31** (2006) 206.

[30] Bocher L., Aguirre M. H., Logvinovich D., Shkabko A., Robert R., Trottmann M. and Weidenkaff A., "$CaMn_{(1-x)}Nb_{(x)}O_3$ ($x <$ or $= 0.08$) perovskite-type phases as promising new high-temperature n-type thermoelectric materials", *Inorg. Chem.*, **47** (2008) 8077, DOI: 10.1021/ic800463s.

[31] Maignan A., Hébert S., Pi L., Pelloquin D., Martin C., Michel C., Hervieu M. and Raveau B., "Perovskite manganites and layered cobaltites: potential materials for thermoelectric applications", *Cryst. Eng.*, **5** (2002) 365, DOI: 10.1016/S1463-0184(02)00048-5.

[32] Fedorov M. I. and Zaitsev V. K., "Thermoelectrics of Transition Metal Silicides", in *Thermoelectrics Handbook: Macro to Nano*, edited by Rowe D. M. (CRC/Taylor & Francis, Boca Raton) 2006, pp. 31/1–31/19.

[33] Bubnova O., Khan Z. U., Malti A., Braun S., Fahlman M., Berggren M. and Crispin X., "Optimization of the thermoelectric figure of merit in the conducting polymer poly(3,4-ethylenedioxythiophene)", *Nat. Mater.*, **10** (2011) 429, DOI: 10.1038/nmat3012.

EPJ Web of Conferences **148**, 00011 (2017) DOI: 10.1051/epjconf/201714800011
LNES 2016

Introduction to wind energy systems

H.-J. Wagner

Ruhr-University Bochum, Energy Systems and Energy Economics
P.O. Box 102148, 44721, Bochum, Germany

Summary. — This article presents the basic concepts of wind energy and deals with the physics and mechanics of operation. It describes the conversion of wind energy into rotation of turbine, and the critical parameters governing the efficiency of this conversion. After that it presents an overview of various parts and components of windmills. The connection to the electrical grid, the world status of wind energy use for electricity production, the cost situation and research and development needs are further aspects which will be considered.

1. – Different wind turbine types: An overview

Today there are various types of wind turbines in operation, (fig. 1 gives an overview). The most common device is the horizontal-axis wind turbine. This turbine consists of only a few aerodynamically optimised rotor blades, which for the purpose of regulation usually can be tumbled about their long axis (Pitch-regulation). Another cheaper way to regulate it, consists in designing the blades in such a way that the air streaming along the blades will go into turbulence at a certain speed (Stall-Regulation). These turbines can deliver power ranging from 10 kW to some MW. The largest turbine on the European

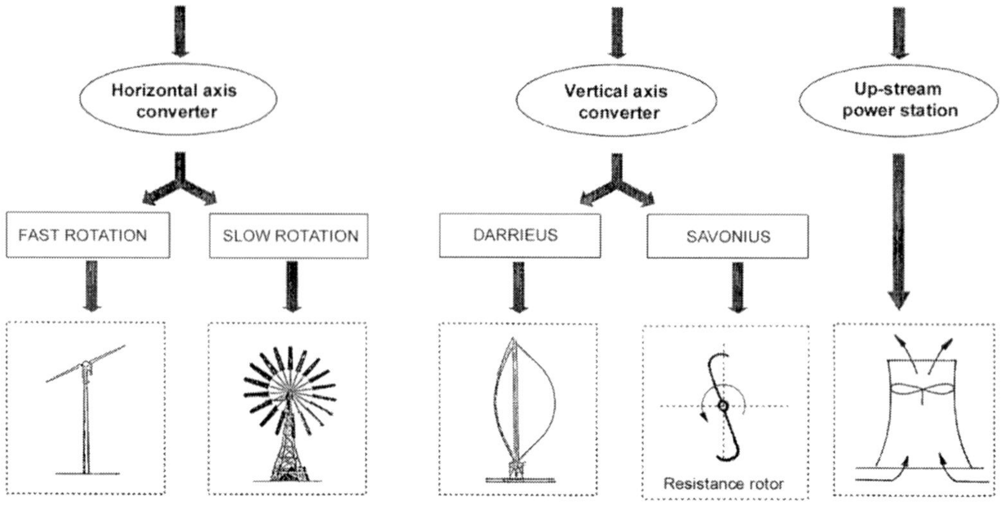

Fig. 1. – Overview of the different types of wind turbines.

market has a power of 8 MW, bigger machines are being tested. The efficiency of this type of turbine is very high. Therefore, it is solely used for electricity generation which needs "high-speed engines" to keep the gear transmission and the generator small and cheap.

Another conventional (older) type of horizontal-axis rotor, is the multiblade wind turbine. It was first built about one hundred years ago. Such wind mills have a high starting torque which makes them suitable for driving mechanical water pumps. The number of rotations is low, and the blades are made from simple sheets with an easy geometry. For pumping water, a rotation regulating system is not necessary, but there is a mechanical safety system installed to protect the turbine against storm damage. By using a so-called wind-sheet in lee direction the rotor is turned in the direction of the wind. In order to increase the number of rotations, this type of turbine had been equipped with aerodynamically more efficient blades facilitating the production of electricity, here the area of a blade is smaller.

The mechanical stability of such "slow-speed turbines" is very high, some have had operation periods of more than fifty years.

A third type of turbine is known as DARRIEUS —a vertical-axis construction. Their advantage is that they do not depend on the direction of the wind. To start, they need the help of a generator working as motor or the help of an SAVONIUS rotor installed on top of the vertical axis. In the nineteen eighties a reasonable number of DARRIEUS-turbines had been installed in California, but a further expansion into the higher-power range and in the European markets has not taken place. One reason may be that they are noisier than horizontal-axis turbines. Another disadvantage is that wind velocity increases significantly with height, making horizontal-axis wheels on towers more economical. Nevertheless, there are some companies producing DARRIEUS-turbines in the very low power range of a few kilowatts for decentralised electricity supply in areas without electrical grids *e.g.* in rural areas of developing countries.

The Savonius rotor is only used for research activities, *e.g.* as a measurement device especially for wind velocity, it is not used for power production. Therefore it will not be discussed in detail in this paper.

The last technique to be dealt with is known as Up-Stream-Power-Station or thermal tower. In principle, it can be regarded as a mix between a wind turbine and a solar collector. In the top of a narrow, high tower is a wind wheel on a vertical axis driven by the rising warm air. A solar collector installed around the foot of the tower heats up the air. The design of the collector is simple; a transparent plastic foil is fixed over several metres on the ground in a circle around the tower. Therefore, the station needs a lot of space and the tower has to be very high. Such a system has a very poor efficiency, only about one percent. World wide there has only been one Up-Stream-Power- Station built so far, it was designed by a German company. For some years it worked satisfactorily at the location of Manzarenas in Spain, but in the mid eighties it was destroyed by bad weather. This station had an electrical power of 20 kW, the tower was about 200 m high, and the collector had a diameter of approximately the same size. A second Up-Stream-Power- Station with an electrical performance of 200 MW was planned in Australia, but not yet realised until. The tower height should be about 1000 m and the diameter of the collector area should be about 7000 m. There has no new Up-Stream-Power-Station been designed and installed so far. Since there has been tremendous technical progress over the last ten years regarding solar farm stations as well as horizontal-axis wind turbines.

2. – Physical basics

2·1. *Energy content of the wind.* – The following section will be used to mathematically explain where the energy in the wind comes from and what factors it depends on.

Power is defined as

$$P = \frac{E}{t} = \frac{1}{2} \cdot A \cdot \rho_a \cdot v^3, \tag{1}$$

with E: kinetic energy, A: area, ρ_a: specific density of the air, v: wind velocity.

Therefore, it is also proportional to the cube of the wind speed, v^3.

From fig. 2, it can be seen that the power output per m^2 of the rotor blade is not linearly proportional to the wind velocity, as proven in the theory above. This means that it is more profitable to place a wind turbine in a location with occasional high winds, than in a location where there is a constant low wind speed. Measurement at different places shows that the distribution of wind velocity over the year could be approximated by a Weibull-equitation. That means that at least about 2/3 of the produced electricity will be earned by the upper third of wind velocity.

From a mechanical point of view, the power density range increases by one thousand for a wind speed change of just 10 m/s, thus producing a construction limit problem.

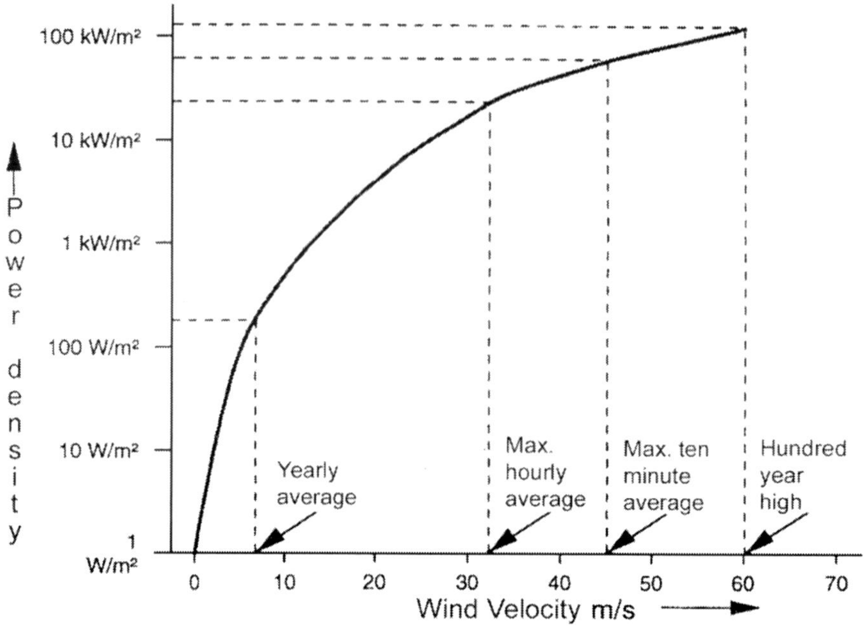

Fig. 2. – Relationship between wind velocity and power output (yearly average valid for Germany)([1]).

Therefore, wind turbines are constructed to harness only the power from wind speeds in the upper regions.

2·2. *Power coefficients.* – There is now the question of how much of the energy in the wind can be transferred to the blade as mechanical energy.

Betz's law and c_p

Betz's law states that you can only convert a maximum of 59% of the kinetic energy in the wind to mechanical energy using a wind turbine. This is because the wind on the back side of the rotor must have a high enough velocity to move away and allow more wind through the plane of the rotor.

The relationship between the power of the rotor blade P_R and the maximum power $P_{R\,\mathrm{max}}$ is given by the power coefficient c_p;

$$P_R = P_1 - P_2 = c_p \cdot P_{R\,\mathrm{max}}. \tag{2}$$

The maximum power coefficient is determined through the ratio v_2/v_1 and setting the

([1]) This figure is taken from [1].

derivation to zero.

$$c_{p\,\max} = \frac{16}{17} = 0.593 \quad \text{with} \quad v_2 = \frac{1}{3} \cdot v_1. \tag{3}$$

Therefore, an ideal turbine will slow down the wind by 2/3 of its original speed (Betz's law).

The issues discussed in the theory can be summed up and related to the design of a wind energy turbine, by the so called Cooking recipe:

"Cooking recipe" for the design of wind turbines.

1. A high aerofoil form ratio leads to a high Tip-speed ratio and therefore, a large power coefficient c_p.

   ⇒ Modern turbines with a good aerodynamic profile rotate quickly.

2. Simple profiles with smaller profile form ratios have a small Tip-speed ratio. Therefore, the area of the rotor radius that is occupied by blades must be increased in order to increase the power coefficient.

   ⇒ Slow rotating turbines have poor aerodynamic profiles and a high number of blades.

3. The profile form ratio and the tip-speed ratio have a considerably greater influence on the power coefficient than the number of blades.

   ⇒ The quality of an aerofoil in respect to a high speed turbine has an inferior significance.

3. – Technical design of wind turbines

3·1. *The design with gearbox.* – The details of a design with gearbox are shown in fig. 3. The main aspect of the classic design is the split shaft system, where the main shaft turns slowly with the rotor blades and the torque is transmitted through a gearbox to the high-speed secondary shaft that drives the few-pole pair generator.

The transmission of torque to the generator is shut off by means of a large disk brake on the main shaft. A mechanical system controls the pitch of the blades, so pitch control can also be used to stop the operation of the turbine in *e.g.* storm conditions. The pitch mechanism is driven by a hydraulic system, with oil as the popular medium. This system needs almost yearly maintenance and constant pressure monitoring, along with the gear box which is lubricated with oil. Due to mechanical losses in the gearbox, the oil will be treated and must be cooled. For constructions without a main brake, each blade has its pitch angle controlled by a small electric motor. This is the standard by most all wind converters.

DOI: 10.1051/epjconf/201714800011

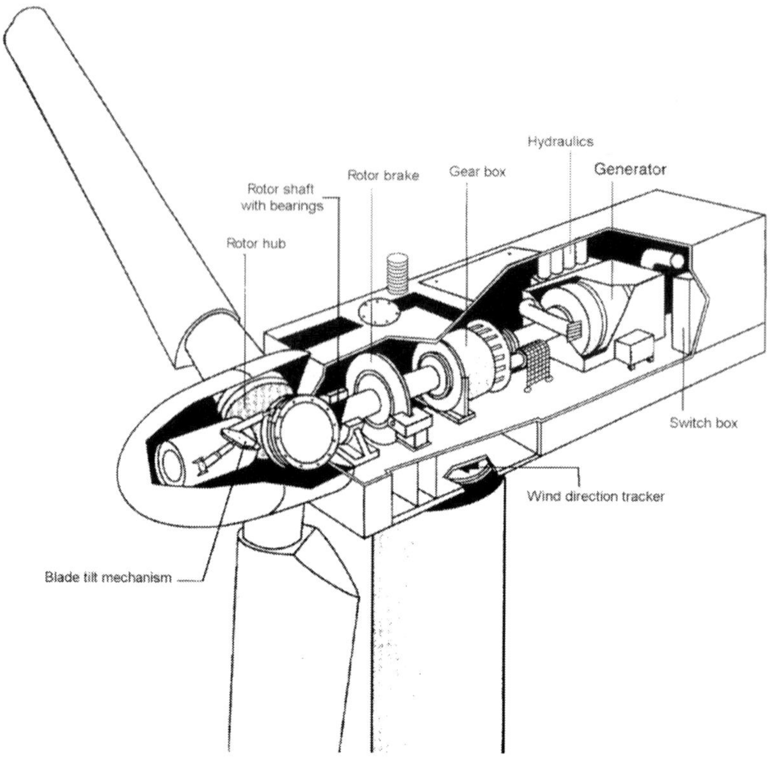

Fig. 3. – The design with gearbox.

To reduce weight generators with permanent magnets was developed. Some producers are equipping their converters with it last years.

In order to reduce vibrations by high speeds in the rotating axe, generators are often designed with more that one pole pair. This allows a reduction in the rotation velocities between the gear box and the generator. The problem with this type of configurations is that the size and weight of the generator increases with the number of pole pairs.

Wind speed and direction measuring apparatus are located at the back of the hub head. A rack-and-pinion mechanism at the join of the hub and the tower, allows the hub to be rotated in to the wind direction, and out of it in storm conditions.

3·2. *The design without gearbox.* – Some companies *e.g.* the German company Enercon, design another turbine type, without gearbox. The scheme of such a turbine is shown in fig. 4, where the main design aspects can be clearly seen.

This design has just one stationary shaft. The rotor blades and the generator are both mounted on this shaft. The generator is in the form of a large spoked wheel with *e.g.* forty-two pole pairs, around the outer circumference and stators mounted on a stationary arm around the wheel. The wheel is fixed to the blade apparatus, so it rotates slowly with the blades. Therefore, there is no need for a gearbox, rotating shafts or a disk brake. This

 DOI: 10.1051/epjconf/201714800011

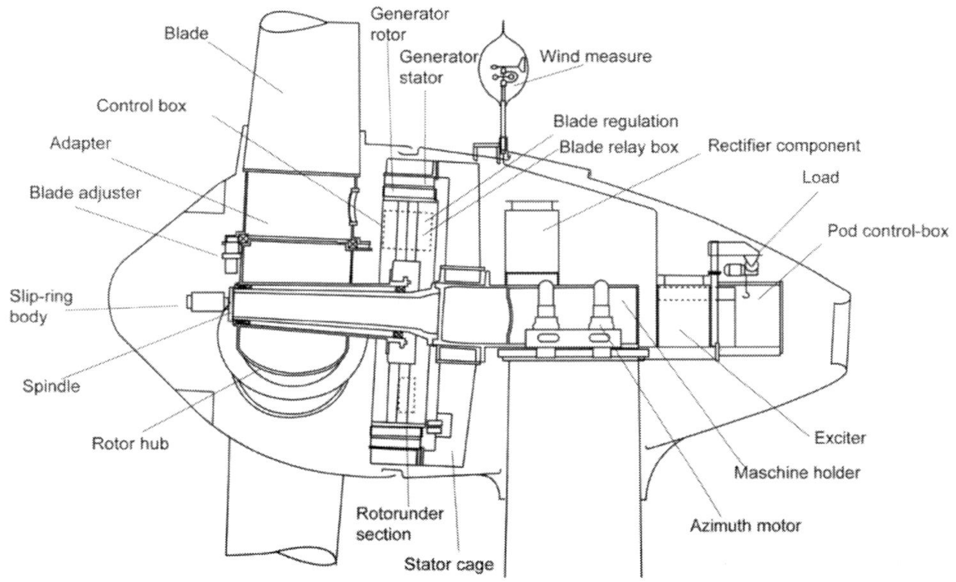

Fig. 4. – The design without gearbox (Enercon E-66).

minimising of mechanical parts simplifies the maintenance and production of the turbine.

The whole system is automated; pitch control and hub direction are controlled by a central computer, which operates the small directional motors.

3·3. *Aspects of design and development.* – There are several critical aspects of a wind turbine that need to be considered in the design phase, to ensure the turbine will be economic and durable.

The tower

In principle, the tower needs to be as tall as possible, because the wind speed increases with height. However, the height is limited by costing issues; an increase in tower height of 10 m costs an extra fifteen thousand Dollars, and a tower height of over 100 m requires an aircraft-warning beacon, which is again so expensive.

Heat energy

Large turbines (> 1 MW), have an average generator efficiency of 98%. Heat is also generated in the mechanical parts of the machine including the bearings and the gear box. This means that around 40 kW of power are lost to the generator heating up during operation. This heat energy needs to be controlled to prevent damage to the machine parts. A large fan system is mounted in on the back side of the hub of a turbine and used to draw cool air through the hub and remove the heat energy emitted during operation.

Control and monitoring

The following aspects of a wind turbine need to be controlled and monitored to ensure effective operation of a wind turbine within the legal limits:

- By large turbines; vibration levels.
- Speed of rotation and the pitch angle, of the rotor blades.
- The natural wind speed and direction.
- The voltage and frequency of the electricity produced.
- The output phase angle compared to the grid phase angle.
- The consistency of the electrical power output.
- The acquisition and storage of electrical signals.
- Signal conversion equipment for the directional motors.
- Rotational speed at night, to reduce the noise levels, because the noise is proportional to the blade-tip speed to the power six.

Mechanical stability

The following forces affect the stability of the mechanical system:

- Gravity.
- Centrifugal forces on the rotor blades.
- Pressure changes on the blade due to the shadow effect the tower creates.
- Stochastic power output of the turbine due to wind energy levels continually changing.
- Resonance of the blades.

Wind direction set-up

A wind turbine can be designed to face in to the wind (windward), or away from it (leeward). A leeward turbine has the advantage of being self orientating, but the disadvantage of the tower disturbing the wind velocity profile, before the wind has reached the plane of the rotor blades. The pressure and speed differences experienced by the blade as it passes the tower, result in stresses on the hub, which need to be alleviated by use of an extra mechanism in the hub to allow the rotor blades to move out of their usual plane of rotation.

TABLE I. – *The technical figures of two different multi-megawatt wind turbines for onshore.*

	Enercon E-182 E3	REpower 3.2 MW
Design	without gearbox	with gearbox
Hub height	80–130 m (onshore)	100–130 m (onshore)
No. of blades	3	3
Rotor speed	6–18 rpm	6.7–12 rpm
Rotor diameter	82 m	114 m
Material of blade	Fibreglass (reinforced epoxy)	Fibreglass (reinforced epoxy)
Blade regulation	Pitch	Pitch
Rated power	3 MW	3.2 MW
Transmission ratio of gearbox	None	approx. 99
Generator	Multi-pole	Asynchronous, few poles
Grid connection	Via frequency converters	Via frequency converters

3˙4. *Technical figures of two modern wind turbines.* – The largest market introduced machine up to the year 2014, is a 8 MW turbine from the company Vestas.

Table I gives an idea of the technical data of wind turbines of the 3 MW class. A design with and without gearbox have been chosen to show the different operation of their generators.

4. – Connection to the electrical grid

The main electrical grid has a constant frequency *e.g.* of 50 Hz or 60 Hz and a constant phase angle. Therefore, a wind turbine must produce electricity with the same constant values in order to be integrated into the main grid.

The input energy of a wind turbine is proportional to the wind speed, but the wind speed is never constant. Each wind speed has a corresponding rotor rotation speed, at which the maximum power is produced. This maximum occurs for different wind speeds at different rates of rotation. However, the rate of rotation must be held constant in order to achieve the required constant output frequency or the wind turbine has to be connected to the grid by doubly fed asynchronous generator or by electronic frequency converters.

A small turbine can be connected directly into the grid network at 0.4 kV. When the wind turbine is integrated into the grid network, there must be no voltage change, voltage oscillation or flicker experienced in the homes on that network branch. The loss of voltage due to resistance in the cabling can be avoided by increasing the diameter of the cables. It is often required that a new network branch is constructed and linked to the transformer, in order to reduce the voltage disturbances. This increases the installation costs of the turbine.

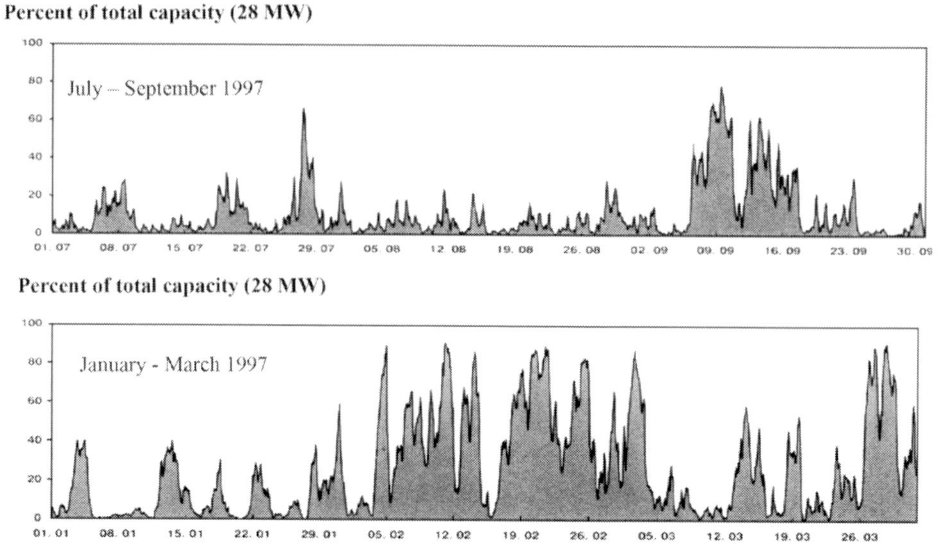

Fig. 5. – Load distribution measurement (Germany).

Megawatt turbines cannot be connected to the grid at the 0.4 kV stage, but have to be connected in at 10–30 kV, which is the usual level of the city electricity share distribution. In remote areas, where a 30 kV connection is not established, the connection must be created and financed by the wind park developers. Wind parks with a lot of Megawatt turbines can also be connected into the electrical grid *e.g.* at the 110 kV level in Germany.

As mentioned earlier the maximum power output is obtained only in few hours during the year. Figure 5 shows a typical load distribution, measured in Germany. With larger wind energy installations in future this uneven distribution leads to the need of higher regulation capacities by conventional power systems.

Grid connection of offshore wind farms poses a technical and economical challenge to wind turbine and grid operators. In the initial phase, the still quite limited capacity of early pilot farms enables using a conventional three-phase AC connection to the onshore grid system which is a well-known technology and inexpensive. This wind farms were also located near to the coast, the energy losses by using AC was therefore limited.

Greater capacities and remote offshore sites make it technically difficult to connect offshore wind farms to the mainland grid by using AC undersea cables. Losses, reactive power production and limited capacity of the sea cables may become important in the future. High-voltage direct current transmission to land could be a solution but it is technically more complicated and more expensive.

At the moment there are some plans for connecting offshore wind farms to the mainland grid. Some lines, AC and DC operated, are already into operation, some more are under realising ore planning. To protect environment on sea several wind farms are connected to one line.

TABLE II. – *World wind power production*(2).

Land/Region	Total installed rated power up to the end of 2015 [GW]
China	145
USA	74
Germany	45
Spain	23
India	25
UK	14
Italy	9
France	10
Canada	11
Brazil	9
Remaining countries	67
Total	432

To connect the offshore wind farm to the onshore grid an internal grid is necessary. The produced power has to be feed to an offshore transformer substation. Wind turbines are connected to it via undersea cables by a voltage of about 30 kV. After stepping-up to the transmission line voltage, the power is conveyed to shore.

5. – Use of wind energy

5·1. *Worldwide status.* – In the nineteen eighties, it was the USA that took the lead in establishing wind farms. They set over 10000 turbines into operation, each generating between 80–200 kW. In Europe it was Denmark that was the main pioneers of wind energy. At the end of twenty-first century, Germany has taken the world lead, producing about a third of the world's wind produced electricity until 2008.

Many governments have began to produce incentive schemes to increase the economic feasibility of wind turbines. Some incentives used include both paying more for wind produced electricity and providing a proportion of the initial construction costs. Governments of industrial countries, or those with a high power consumption, are eager to promote wind energy, because it is environmentally clean and sustainable and limits the need for fossil fuel usage (table II).

The world total rated power was by the end of the year 2015 about 432 GW. This is about 1.8% of the worldwide installed capacity of power stations (water, coal, naturalgas, nuclear).

(2) This statistic is taken from [2].

TABLE III. – (*a*) *The typical investment costs of a* 2 MW *onshore wind power station in Germany* (*Amesdorf*) (*Status: 2013*)([3]). (*b*) *The typical operating costs of a* 2 MW *onshore wind power station in Germany* (*Wellen*) (*Status: 2013*)([3]).

(a)

Investment plan	Costs [€/kW]	
Hub height	< 120 m	> 120 m
Wind power station, transport, installation	1150	1340
Foundation	70	
Grid connection	70	
Site development (lanes)	40	
Planning, environmental measures, concession, others	190	
Total	1520	1710

(b)

Operating costs: 5.1 ct/kWh (Average over 20 years operating time)	
Service, reparation, others	50%
Rent	20%
Management (technical and business)	20%
Reserve for unforeseen events	5%
Insurance	5%

It should be taken into account that wind energy have a lower capacity factor than fossil and nuclear power stations (around 1) with around 0.3 to 0.4 for onshore wind power and around 0.5 for offshore wind power.

5˙2. *Offshore wind status.* – Due to high wind potential at Germany's coastline, wind energy should play a decisive role. Therefore, Germany's government plans to install around 6.5 GW of offshore wind parks until 2020 and 15 GW until 2030. By the end of the year 2015, the wind power installed capacity in Germany for offshore was around 3.4 GW and for onshore around 42 GW [3].

5˙3. *Investment and operation costs.* – The costs involved in installing a wind turbine vary depending on the design, size and chosen location of the new turbine. The infrastructure costs can be minimised by constructing wind parks, where a number of new turbines are installed on the same sight. An example of the investment costs for a wind park at land in Germany is shown in tables III(a) and III(b).

The money invested in an average wind park on land depreciates over about a ten year period. During this period, the set-up and installation costs are high, along with the

([3]) These data are taken from WindGuard GmbH, Germany.

DOI: 10.1051/epjconf/201714800011

loan repayments and insurance costs. After this period, the costs then decrease. Over the next ten years, a financial return can then be made on the investment, when the price for the electricity per kilowatt hour is set at a high value by the government. This means it is economic to ensure the durability of the installed turbines, so maintenance costs after the ten year period are kept low and the investors can receive a good return on their investment.

The above example of investment costs for wind parks on land in Germany is non-transferable to other countries. Especially the employee's wages and the structure of financing wind farms are quite different. Nevertheless power generating by wind turbines in developed countries with existing electrical grid and sufficient installed power, costs three to four times in comparison to power generation by conventional power stations.

Wind parks on sea need higher specific investments, although the wind offer is better on sea than on land. The costs of produced electricity are higher by an offshore wind park (status 2012). The expectation is that the cost could be reducing for new offshore wind parks in next years. A lot of countries are supporting the installation of wind converters by supporting programs.

5˙4. *Environmental aspects.* – Wind energy is a renewable energy source; therefore it holds many advantages over the fossil fuels, which have diminishing reserves. Wind energy is clean in regard to toxic emissions. Therefore, it does not add to Global warming or Acid rain problems.

The wind turbines can affect the environment in aesthetic and human intrusive ways. This is because they must be sited in prominent locations and through the nature the rotation of their blades; they produce optical distortions, *i.e.* flickering shadows, and a humming noise. The land required for the sighting of a wind park can be considered large, if all the access routes are also taken into consideration. However, they very rarely require the resettlement of communities, which is a problem associated with *e.g.* large Hydro-Electric-Power stations. The danger to birds of the rotating blades has been questioned, but it has been found that the birds change their flight paths to avoid the blades. It has also be questioned whether the reduced wind speed at ground level, affects the growth of flora. This is answered by the observation that many wind parks have animals grazing between the turbines.

5˙5. *Life cycle analysis.* – Life Cycle Assessments (LCA) is an important tool for industry and policy makers, used to determine the actual emissions of a product or technology throughout its whole life cycle. In case of energy production systems or power plants, analysis of energy required to produce the materials and processes; emissions resulting from various processes for materials production and processes resulting into their Cumulated Energy Demand (CED) and Global Warming Potential (GWP) become important parameters when making decisions on further research, development and deployment of any technology. Results and characteristics of wind energy LCA were for the first time analyzed in the German offshore wind farm alpha ventus. Alpha ventus is a test field in the North Sea for 5 MW wind energy converters in a water depth of 30 m.

Results shown that within less than one year the energetic expenditure as well as the greenhouse gas emission of the entire life cycle of alpha ventus is amortized [4].

6. – Research and development needs

The field of wind converter has been connected with high technical development activities for many years. The first report of the project "Wind Energy Thematic Network" [4] founded by the European Commission described in the year 2012 some aspects, which are also relevant in:

1. *Environmental and social impacts* (*e.g.* enhancing local incentives by developing participation models):
   - methods to integrate wind turbines visually in to the landscape,
   - reduction of noise impacts,
   - mitigating impacts on bird populations, habitats and flight paths,
   - turbine design regarding life cycle analysis,
   - analyse social effects like local employment, investment, taxes etc.
2. *Wind turbine and component design issues* (*e.g.* basic research in aerodynamics, structural dynamics, structural design and control):
   - new materials with higher strength like carbon fibre for the blades,
   - feasibility studies of new wind turbines concepts and innovations,
   - integration of demand site requirements in the design of turbine, *e.g.* electrical control system interaction with grid requirement.
3. *Testing, standardisation, and certification* (*e.g.* common accepted certification procedures for wind turbines and wind farms):
   - identification of standards lacking, and initiation of appropriate actions for new standards,
   - standards for service and maintenance concepts of offshore wind converters.
4. *Grid integration, energy systems and resource prediction* (*e.g.* forecast of wind resource):
   - development of scenarios for redesigning the grid system with high wind penetration,
   - increasing both: power quality and consistency,
   - energy management and storage systems for stand alone applications.
5. *Operation and maintenance* (*e.g.* advanced condition monitoring):
   - development of early failure detection and condition systems,
   - development in preventative maintenance,

TABLE IV. – *Technical developments in Wind Turbines* [5].

Generators	– Copper? Permanent Magnet? – Rare earths? – Superconductors?
Rotor Blades	– Cost reduction? – Utilization? – Legal situation? – Active control elements in rotor blades?
Operations Monitoring	– Drive train monitoring (CMS) for onshore plants? – Tower- and foundations vibrations – Number of Sensors, also Onshore? – Periodic Inspections- Scope, Number?
Technical Uncertainties	– Handling of new developments?
Offshore Foundations technologies	– Steel? Concrete? – Gravity Foundations? – Environment, under water noise protection during the piling for the foundation?

– standardisation of components for easy replacement,
– certification of service and maintenance concepts,
– cold and icing climates resource assessment.

6. *Offshore wind technology* (*e.g.* research into the control and efficiency of very large wind farms and more cost effective foundations, transport and installation techniques):

   – monitoring of environmental impacts (effects on birds, effects of noise and vibration on marine life especially by installation of wind converter etc.),
   – development of deep water foundation structures (fundamental wind turbine design research),
   – offshore meteorology,
   – special designs of systems and components for transportation, erection, access and maintenance of offshore wind turbines,
   – investigate the use of energy storage,
   – improve corrosion protecting systems regarding the offshore conditions.

In addition to these aspects there have been done a lot of R&D needs by operating and market introduced wind turbines, *e.g.*:

– better state of knowledge of the dynamic forces on the drive train,

– improvement of the availability of gear boxes,

– optimisation of control units and control systems,

– development of central lubrication systems.

From a technical perspective several configurations for wind technologies are under research and development in Germany (table IV).

7. – Outlook

The wind energy market has grown because of the environmental advantages of harnessing a clean and unexhaustable energy source and because of the economic incentives supplied by several governments. However, energy is required from other generation methods during the building phase of a new turbine, so in this period, Greenhouse gases and air pollution will be added to. If the life cycle of a wind turbine is looked at, more pollutants are saved on during operation, than are emitted during the building phase.

A wind turbine is not a self-sustainable power station. This means that back-up power generation is needed at the times when the turbine is inoperative. This back-up is nowadays supplied by the established fossil fuel power stations. If the number of wind turbines increases, the need for extra investment in the back-up generation systems will rise, in order to maintain a stable electricity grid system. The challenges in this case are that existing power stations will deliver less electricity than before. Under this condition the electricity generation prices for new power stations are increasing due to smaller load duration time than before.

However, wind energy is still one of the most important renewable energy resources for the future, because it can be harnessed in a clean and inexhaustible manner, through the application of technically advanced and efficient machinery. These additional investments will need in some cases subsidies. For more information especially in respects to technological and ecological aspects (life cycle aspects) see [4].

REFERENCES

[1] Kleemann M. and Melib M., *Regenerative Energiequellen*, 2.Auflage (Springer-Verlag, Heidelberg) 1993.
[2] `www.CWEC.de`.
[3] Ender C., "Wind Energy Use in Germany", DEWI Magazin 02/2016.
[4] Wagner H.-J. and Mathur J., *Introduction to Wind Energy Systems – Basics, Technology and Operation*, second edition (Springer-Verlag, Berlin) 2013, ISBN 978-3-642-32975-3.
[5] Wagner Hermann-Josef, "Erneuerbare Energien Wind und Sonne - Wo geht es technisch hin?", VGB PowerTech 1/2, 2015.

Geothermal energy

A. Manzella

Institute of Geosciences and Earth Resources - Pisa, Italy

Summary. — Geothermal technologies use renewable energy resources to generate electricity and direct use of heat while producing very low levels of greenhouse-gas (GHG) emissions. Geothermal energy is the thermal energy stored in the underground, including any contained fluid, which is available for extraction and conversion into energy products. Electricity generation, which nowadays produces 73.7 TWh (12.7 GW of capacity) worldwide, usually requires geothermal resources temperatures of over 100 °C. For heating, geothermal resources spanning a wider range of temperatures can be used in applications such as space and district heating (and cooling, with proper technology), spa and swimming pool heating, greenhouse and soil heating, aquaculture pond heating, industrial process heating and snow melting. Produced geothermal heat in the world accounts to 164.6 TWh, with a capacity of 70.9 GW. Geothermal technology, which has focused for decades on extracting naturally heated steam or hot water from natural hydrothermal reservoirs, is developing to more advanced techniques to exploit the heat also where underground fluids are scarce and to use the Earth as a potential energy battery, by storing heat. The success of the research will enable energy recovery and utilization from a much larger fraction of the accessible thermal energy in the Earth's crust.

1. – Introduction

Geothermal technologies use the heat contained in the ground. In its broad term the geothermal energy is the thermal energy contained in our planet, the Earth. In this document we will restrict our attention to the practical use for mankind, *i.e.* the heat that can be extracted from a depth up to about 10 km, which is the maximum depth reached by human drilling. In this thin portion of Earth heat is inexhaustible, since it moves from the Earth's interior towards the surface, where it dissipates. Part of this heat derives from the slow cooling from primordial heat, involved in the Earth's formation, and part derives from the decay of long-living radioactive isotopes contained by crustal rock minerals. The most abundant isotopes having a half-life comparable to the age of the Earth, so that they can still be significant heat producers, are ^{40}K, ^{232}Th, ^{235}U and ^{238}U. The amount of heat that is released to the space from the interior through a unit area in a unit of time, the Earth's *heat flow*, varies from place to place and with time, depending on the geological and therefore physical condition of the underground. On average, the heat flow from the continental crust is 57 mW/m^2 and from the oceanic crust is 99 mW/m^2 [1]. Our planet is a dynamic system, with its inner changes, mostly slow from the perspective of human time, but producing very quick changes close to the surface. These are not only earthquakes and volcanic eruptions, very well-known to humans. The interaction of inner, geological processes with the physical processes just above the ground surface originates the complex system characterizing the geothermal systems. The hottest areas of the Earth's surface are near tectonic plate boundaries, which are often associated also to volcanoes and seismic activity and where hot magmatic bodies are present within or close to our reference 10 km depth. The heat flow at the Earth's surface therefore derives from the radiogenic decay within the upper crust, the heat generated in the most recent magmatic episode, and the heat coming from the mantle and inner Earth's structures. In continental crust, where the isotope heating is larger since granitic rocks rich of long-living radioactive isotopes are common, the heat flow is highest in areas having experienced "recent" (less than 65 million year) magmatic activity, whereas it decreases to 46 mW/m^2 in crust older than 800 million years [1]. In oceanic crust the concentration of radiogenic heating is negligible due to the rock composition (basaltic), but the crust thickness is smaller than in continents and the heat flow largely derives from heat flowing from the mantle. Most plate boundaries are below sea level and the young oceanic crust has very high heat flow values. Geothermal exploitation is focused so far to continental areas, where energy is requested.

To extract energy from the underground we mostly use the water as carrier of heat. As the crust is highly fractured and thus permeable to fluids, surface water, essentially rainwater, penetrates at depth and exchanges heat with the rocks. Two main forms of heat transfer occur within the crust: conduction and convection. Where rocks are very fractured and circulating fluids are abundant, the resulting convective heat transfer is very efficient and can be easily exploited by drilling wells and discharge the hot fluids to the surface. In these convective systems, named hydrothermal resources, the aquifers represent the geothermal reservoir. Occasionally, in areas of very high heat flow, the fluid has high temperature (up to above 300 °C) and, depending on the pressure, can be vapor (steam) or water.

Warm and hot fluids can be extracted from the underground in a wide range of temperature and discharge rate, and used directly for their heat content or to produce electric power. Even the modest temperatures found at shallower depths can be used to extract or store heat by means of ground source heat pumps, that are nowadays a widespread application for geothermal energy.

DOI: 10.1051/epjconf/201714800012

Geothermal technologies described in the following sections, while producing very low levels of greenhouse-gas (GHG) emissions, have an important role to play in realizing targets in energy security, economic development and mitigating climate change. Geothermal resources, as opposed to hydrocarbon ones, are generally *renewable* since the circulation of heat and fluid is continuous. There is a constant terrestrial heat flow to the surface, then to the atmosphere from the immense heat stored within the Earth, and fluid enters the reservoir from the recharge zones or injected in the subsurface through injection wells (in industrial plants). Heat can be extracted at different rates. To guarantee a *sustainable* use of geothermal energy, the rate of consumption should not exceed the rate of generation, so that the heat removed from the resource is replaced on a similar time scale. Geothermal plants typically develop below a certain level of energy production.

Geothermal typically provides base-load generation, since it is generally immune from weather and seasonal variation, therefore producing almost constantly and distinguishing it from several other renewable technologies that produce variable power or heat with time.

2. – Utilization of geothermal heat

In direct use geothermal energy is employed directly as heat without further conversion into other types of energy. On 2015 at the World Geothermal Congress 83 countries were reported to use geothermal heat for direct uses. The overall capacity of 70.3 GW was installed on 2015 and produced in total 58.7 EJ of thermal energy [2] (fig. 1).

Direct use of geothermal heat and water dates back thousands of years and it continues today. The Romans, Chinese, and Native Americans left clear proofs that they used geothermal water for bathing, cooking and for therapeutic purposes. Where accessible, natural hot waters have been used for space heating in cold areas and seasons. Today geothermal water is used for many different applications depending on the temperature of the water, including electricity generation, which is the most important form of utilization of high-temperature geothermal resources ($> 120\,°\text{C}$). Known uses of geothermal heat range from fish farming and soil heating at low temperatures, through space heating and drying at intermediate temperatures, to industrial processing at high temperatures (they have a wide temperature range).

Heat demand represents a significant share of final energy consumption for space heating, especially in cold countries, agricultural and industrial processes. Geothermal heat production systems may meet the demand simply providing fluids at the required temperature. For example, the water circulating through the network of pipes, beams and panels of the heating systems of a building may be kept constant in the summer by exchanging heat with groundwater fluids at a temperature of 18 °C. If groundwater fluid temperature of 30 °C is discharged, it can be used to pasteurize milk in a cheese factory. This passive or "free" heating and cooling (H&C) is one of the most environmental friendly forms of H&C available, and the base of most of the direct use of geothermal heat till recently. But passive use has a limitation: it must adapt to the temperature of natural groundwater available, which is not always what we need. Thanks to improvements of

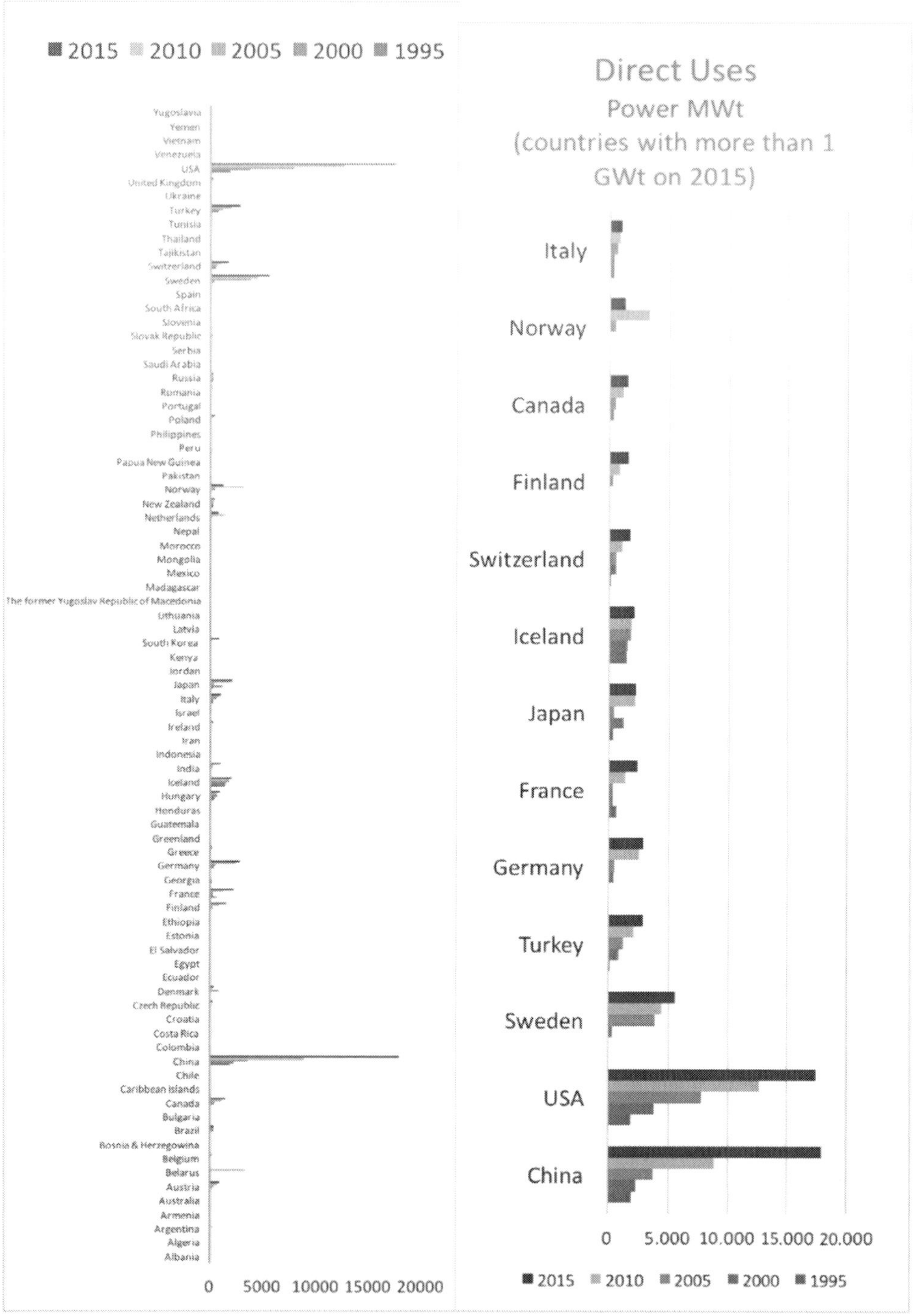

Fig. 1. – Thermal installed capacity trend for 83 countries (left) and for the 13 countries with capacity larger than 1 GW on 2015 (right).

absorption machines and heat pumps, which are readily available on the market and efficiently able to heat and cool the fluids, geothermal applications are becoming more and more flexible.

Geothermal heat pump systems for the moment are a reference market mostly dedicated to space heating and cooling, and exploits groundwater or groundcoupled temperatures at shallow depth (10–200 m depth). Since they do not require any particular thermal anomaly, geothermal heat pumps represent a geothermal heat production technology which is suitable even for regions with altogether ordinary geothermal conditions. Thanks to them, nowadays direct use of geothermal heat can be done anywhere in the world, and many countries which are not specially known for geothermal resources have been ranking high in geothermal direct utilization due to the use of geothermal heat pumps. Since this technology has a large potential of expanding its application and may be combined with other geothermal technologies, we describe it in some detail.

2˙1. *Geothermal heat pump technology.* – To heat or cool a space we need to transfer heat. To this aim, we may use the thermal energy stored in the ground at shallow depth, usually 50–100 m to a maximum of 250 m. Below a depth of a few meters the temperature is not affected by the seasonal change of air temperature and is essentially constant and equal to the mean annual air temperature. The thermal energy of the ground can be exploited by using heat pumps, which convert the low-temperature geothermal energy to thermal energy at a higher temperature exploiting the physical property of fluids to absorb and release heat when they vaporize or condense, respectively. They work like a refrigerator, moving heat from a space (to keep it cool) and discharging heat at higher temperature. They are made of a closed circuit where a refrigerant fluid is circulated, usually water or other kind of fluids not dangerous for the environment. Depending on temperature and pressure condition the refrigerant fluid is in the liquid or vapor state, and is compressed or expanded by using an external energy, usually electric, to move the heat. Common air conditioning systems work this way and exchange heat with the external air.

In the Ground Source Heat Pumps (GSHP) the refrigerant fluid exchange heat with the ground, taking advantage of the fact that the ground has a stable temperature throughout the year. In winter, when the external air is cold, the ground has a higher temperature and the heat exchange is more efficient. With proper technology the process may be reversed, and the heat pump may also work as a refrigerator in summer, using the cool temperature of the ground to chill the space. The main component added is the underground heat exchanger, through which the heat is absorbed or released to the ground.

There are two main kinds of GSHP: the open and close loop systems (fig. 2). Open systems physically abstract water from the source. Groundwater is abstracted (usually pumped) from springs, drilled boreholes or flooded mines. In closed loop systems a carrier fluid circulates in a closed loop of pipe buried underground and is not in direct contact with the ground or rock. The latter is often called subsurface heat exchanger or *geoexchanger*.

 DOI: 10.1051/epjconf/201714800012

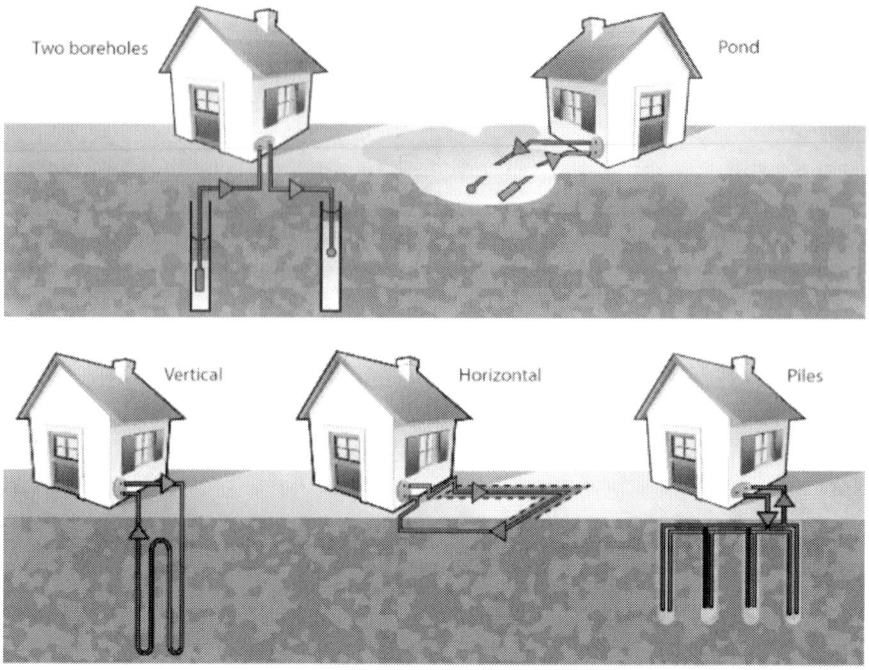

Fig. 2. – Open (top) and closed (bottom) loop GSHP systems.

The *open loop systems* are made of a heat exchanger (in the building to be heated/ cooled), an abstraction well, an injection well (fig. 3). In some cases the abstracted water is not re-injected in the ground but released at the surface (in sewage systems, for example), thus avoiding the cost of a second borehole. This practice is recommended only when there is no risk of reducing the groundwater in the aquifer, and should be done with the proper care for avoiding environmental impact. Re-injection of abstracted water is also regulated by local laws, which defines limitations of use.

The amount of water to abstract, the rate of abstraction and the distance between the two wells must be properly planned. They depend on the heat demand and on the thermal and hydraulic properties of the ground. Accurate planning avoids to cool (or heat in summer) the ground at a higher rate than natural equilibrium allows, therefore guaranteeing the sustainability of the systems and the thermal balance in the aquifer. The balance is enhanced when the system works in reverse mode, *i.e.* heating in winter (chilling the ground) and cooling in summer (heating the ground). Abstraction and injection wells should be far enough to avoid thermal short circuiting between the two during the installation's life time, commonly 20–30 years.

The open systems have the advantage of using natural water, and may be used also for "free" (natural) cooling or for drinking, watering depending on its quality. Moreover, they absorb more heat from the ground than a geoexchanger of the same size, and for equal heat demand they cost less since they require less drilling.

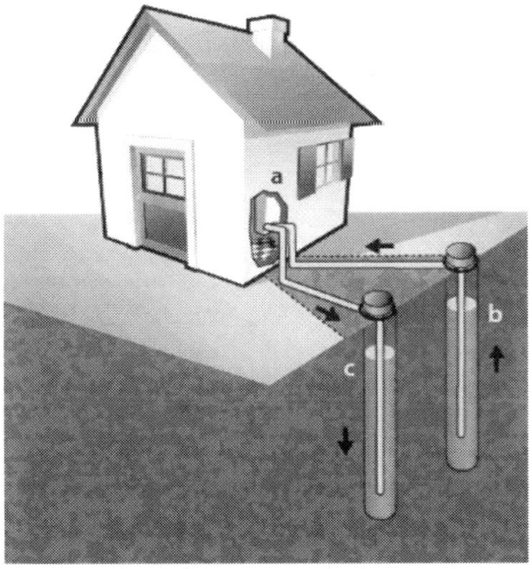

Fig. 3. – Open loop GSHP system in detail: a) heat exchanger; b) abstraction well; c) injection well.

The *closed loop or Ground Coupled Heat Pump* (GCHP) systems are made of closed loop tubing that can be installed horizontally as a loop field in trenches, vertically in wells or as baskets (figs. 4, 5 and 6). The loop is typically made of high-density polyethylene pipe and contains a mixture of water and anti-freeze. The size of the loop field depends on the soil type and moisture content, the average ground temperature and the heat loss or gain characteristics of the building being conditioned. Horizontal loops can be used wherever there is sufficient surface area available for their installation, and are more rarely installed for space heating and cooling of buildings than vertical borehole heat

Fig. 4. – Closed loop GSHP, horizontal linear system.

 DOI: 10.1051/epjconf/201714800012

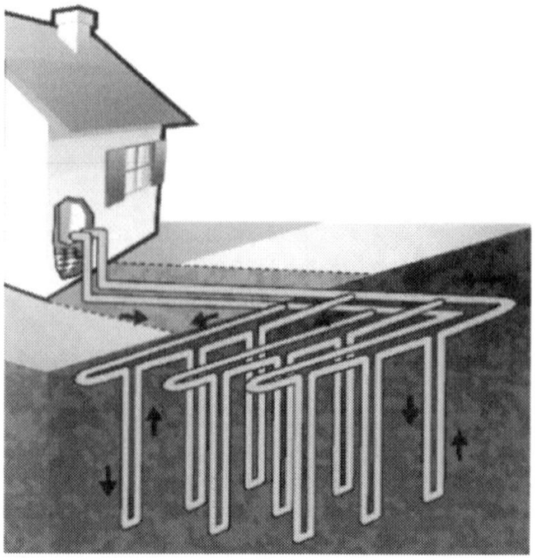

Fig. 5. – Closed loop GSHP, vertical system.

exchangers. Vertical exchanger commonly consist of one or several U-pipes or coaxial pipes, more frequently used for deeper boreholes. Bore-holes are spaced at least 5–6 m apart and the depth depends on ground and building characteristics. For example, a house needing 10 kW of heating capacity might need three boreholes 80–110 m deep in sedimentary rocks. Heat exchanger pipe systems can be integrated directly into the concrete foundations of buildings and other constructions for heating and cooling and are called heat piles. If properly designed and integrated into a combined heating and cooling system right from the beginning, these systems can be a useful part of modern low-energy, low CO_2-emission buildings and constructions.

Fig. 6. – Closed loop GSHP, basket system.

The net thermal efficiency of a heat pump should take into account the efficiency of electricity generation and transmission, typically about 30%. Since a heat pump moves 3 to 5 times more heat energy than the electric energy it consumes, the total energy output is much greater than the electrical input. This results in net thermal efficiencies greater than 300% as compared to radiant electric heat being 100% efficient. Traditional combustion furnaces and electric heaters can never exceed 100% efficiency.

Geothermal heat pumps can reduce energy consumption —and corresponding air pollution emissions— up to 44% compared to air source heat pumps and up to 72% compared to electric resistance heating with standard air-conditioning equipment. The performance of heat pumps is usually expressed as Coefficient of Performance (COP), which is the ratio of heating output or heat removal to electricity input. Cooling performance is typically expressed as the Energy Efficiency Ratio (EER) while heating and cooling performance of reversal heat pumps is typically expressed as the Seasonal Performance Factor (SPF), the mean of COP and EER.

Performance is influenced by all components of the installed system, including the soil conditions, the ground-coupled heat exchanger, the heat pump appliance, and the building distribution, but is largely determined by the difference between the input (ground) temperature and the output (room) temperature. GSHP on the market today have standard COP ranging from 2.4 to 5.0 and EER ranging from 10.6 to 30 [3].

Geothermal heat pump technology is used for space heating and cooling of single building, small and large, and in district heating systems. Ground source heat pumps are characterized by high capital costs and low operational costs compared to other conventional heating source. Since the overall economic benefit depends primarily on the relative costs of electricity and fuels, which are highly variable over time and across the world, a constant and overall estimate is not possible. Based on recent prices, ground source heat pumps have lower operational costs than any other conventional heating source except natural gas and almost everywhere in the world, excluding a handful of countries where natural gas is exceptionally cheap, or where electricity is exceptionally expensive. Capital costs, which may amount to 10–15 thousands of US$ for a single home system, are known to benefit from economies of scale so they are more cost-effective for larger commercial buildings and harsher climates. The cost of installation, which can be two to five times that of a conventional heating system in most residential applications, is affected by the size of living area, the home's age, insulation characteristics, the geology of the area, and location of the property. Although ground source heat pumps are recognized as one of the most efficient heating and cooling systems on the market, they are known and advertised only in some countries, and seldom benefit from tariffs and incentives helping to sustain the capital cost and based on metered readings on an annual basis for 1 or 2 decades.

Technological frontiers in the geothermal heat pump sector are dedicated to the optimization of seasonal thermal energy storage and interseasonal heat transfer. The heat captured and stored in thermal banks in the summer can be retrieved efficiently in the winter. The Underground Thermal Energy Storage (UTES) concept goes beyond the geothermal energy and consider the storage of thermal energy produced by solar source

DOI: 10.1051/epjconf/201714800012

Fig. 7. – Sketch of Borehole Thermal Energy Storage, BTES, in the cooling phase of summer (left) and the heating phase of winter (right).

or waste heat from industrial processes, for example. UTES can be closed (Borehole Thermal Energy Storage, BTES, fig. 7) or open systems (Aquifer Thermal Energy Storage, ATES, fig. 8), as for GSHP technology. Heat storage efficiency increases with scale, so this advantage is most significant in commercial or district heating systems.

Other technological frontiers refer to optimization of process technology of heat pumps, such as for the absorption heat pump, and the combination of geothermal and other renewable energy technologies for heating and cooling.

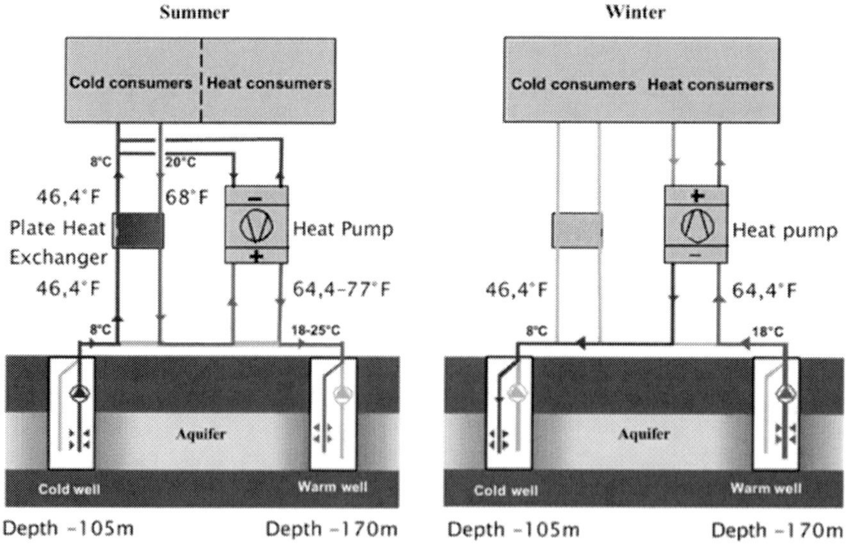

Fig. 8. – Sketch of Aquifer Thermal Energy Storage, ATES, in the cooling phase of summer (left) and the heating phase of winter (right).

2·2. *Technology for direct use of heat.* – A variety of processes requires a temperature comparable to that of available geothermal fluids. The warm water circulating in pipes and appliances for providing heat to these processes is conventionally heated with a boiler, using gas or other fossil fuels as energy source. When geothermal fluids are used, they can be directly circulated in the heating system or, more frequently, clean water is heated through a heat exchanger and then circulated, while geothermal fluids are re-injected in the subsurface.

The use of geothermal heat not only increases the share of green energy and reduces GHG emissions, but on the long term it also contributes to the energy efficiency of the plant since it forces the dimensioning of the heating system to the true temperature level required by the process. Most heating systems are dimensioned for very high temperature waters, easily obtainable by boilers but not with geothermal. It has not been easy or convenient to adapt an existing heating system to a geothermal exchange system, and most of reported geothermal direct use plants were designed as geothermal since the beginning. Thanks to heat pump technology we have now much more flexibility.

The main types of direct applications of geothermal energy are space heating (including heat pumps described before), bathing and swimming (including balneology), agricultural (greenhouses and soil heating), industry, and aquaculture (mainly fish farming). In Europe the main share of installed thermal potential is for geothermal heat pump system (68%) and district heating (15%), while balneology, aquaculture and deicing amount to 11% and industrial and agricultural processes to 6%. There are many other types of utilization, on a much smaller scale, some of which are unusual and seldom reported [2].

Space and district heating and cooling has its most important expression in Iceland, where over 98% of space heating is provided by geothermal. It is also widely distributed in the East European (Pannonian) countries, as well as in the United States, China, Japan, France (Paris basin).

Geothermal district heating (GDH) systems are capital intensive. The main costs are initial investment costs for production and injection wells, and down-hole pumps. These costs add to the high cost of conventional district systems, for transmission pumps, pipelines and distribution networks, monitoring and control equipment, peaking stations and storage tanks. Operating expenses, however, are comparatively lower than in conventional systems, and consist of pumping power, system maintenance, control and management. An important benefit is achieved by combining heating and cooling in areas where the climate permits it, since the load factor would be higher than the factor for heating alone, and the unit energy price would consequently improve. The distribution network is particularly expensive, and a crucial factor in estimating the initial cost of the system is the thermal load density, which is the heating and cooling demand divided by the ground area of the district. A high heating and cooling demand density determines the economic feasibility of a geothermal district heating project.

In Europe there is an increasing interest in GDH systems. Of the about 5000 district heating systems of Europe, 237 are GDH (197 in EU-28) and cover about the 10% of heat market, with a total thermal capacity of 4.3 GW and about 12883 GWh of production [3].

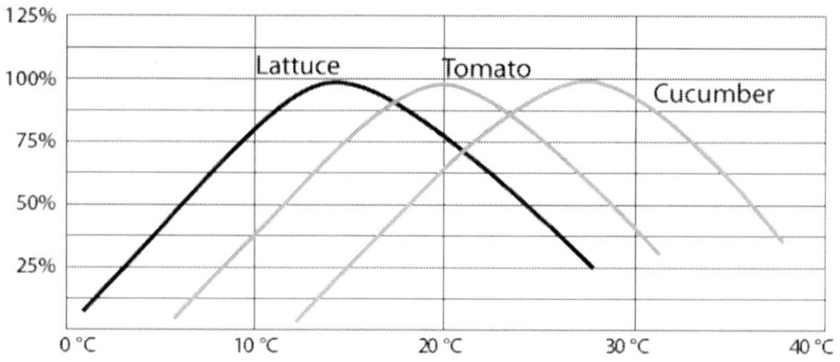

Fig. 9. – Productive trend with temperature of common vegetables.

The *agricultural applications* of geothermal fluids consist of the plant growing temperature control in open fields and greenhouses. Thermal water can be used in *open-field agriculture* to heat the soil and, sometime, to irrigate it where ponds and rivers are not available. Soil heating is provided by burying thin pipelines where warm fluids are circulated. The greatest drawback in this heating is the cost, and this technique results economic only in few cases. Heating the soil in buried pipelines without the irrigation system could also decrease the heat conductivity of the soil, because of the drop in humidity around the pipes, and consequent thermal insulation. The best solution seems to be that of combining soil heating and irrigation. The chemical composition of the geothermal waters used in irrigation must be monitored carefully to avoid adverse effects on the plants. The temperature control in open-field agriculture prevents any damage ensuing from low environmental temperatures, extends the growing season, increases plant growth, and boosts production. It also sterilizes the soil, increasing productivity.

The most common application of geothermal energy in agriculture is, however, in *greenhouse heating*, which has been developed on a large scale in many countries. The greenhouse heating extends the cultivation of vegetables and flowers out-of-season or in an unnatural climate, and can now draw on a widely experimented technology. Various solutions are available for achieving optimum growth conditions, based on the optimum growth temperature of each plant (fig. 9), and on the quantity of light, on the CO_2 concentration in the greenhouse environment, on the humidity of the soil and air, and on air movement. Greenhouse heating can be accomplished by forced circulation of air in heat exchangers, hot-water circulating pipes or ducts located in or on the floor, finned units located along the walls and under benches, or a combination of these methods.

Operating costs of greenhouse heating, which in some cases account for 35% of the product costs (vegetables, flowers, house-plants and tree seedlings), can be considerably reduced by geothermal heat. This is the case in particular when geothermal fluids naturally flow at the surface and abstraction pumps are not required. Geothermal brines often require treatment, since their chemical composition may produce deposits in the pipes, pumps and heat exchanger plates.

DOI: 10.1051/epjconf/201714800012

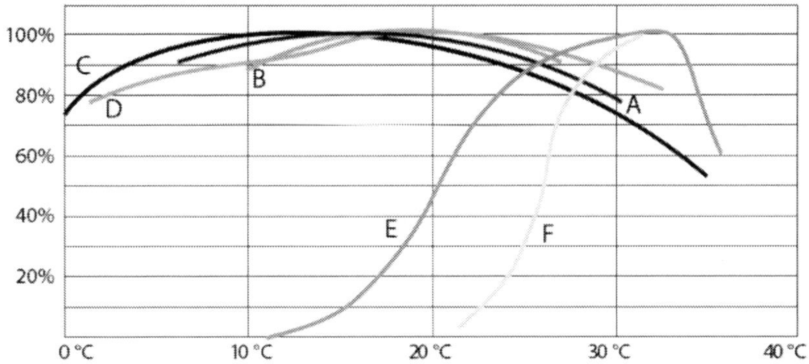

Fig. 10. – Productive trend with temperature of different animal species. A = chicken, egg production; B = poultry, weight increase; C = Jersey cow, milk production; D = pork, weight increase; E = shrimps, weight increase; F = cat-fish, weight increase.

This technology has taken maximum advantage from heat pump technology, and shallow groundwater or ground energy is more and more used. In the Netherlands, for examples, there has been a major boost of local tomato production thanks to the geothermal technology.

Farm animals and aquatic species, like vegetables and plants, benefit in quality and quantity from the optimum conditioning of their environmental temperature (fig. 10). Geothermal heat is used profitably in animal husbandry. The energy required to heat a breeding installation is about 50% of that required for a greenhouse of the same surface area, so a cascade utilization could be adopted. Breeding in a temperature-controlled environment improves farm animal health, and the hot fluids can also be used to clean, sanitize and dry the animal shelters and waste products. Aquaculture, the controlled breeding of aquatic forms of life, is the most common application of geothermal animal husbandry, due to an increasing market demand and the large importance of the breeding temperatures for aquatic species with respect to land species, as can be seen in fig. 10. By maintaining an optimum temperature artificially we can breed more exotic species, improve production and even, in some cases, double the reproductive cycle. The species that are typically raised are carp, catfish, bass, tilapia, mullet, eels, salmon, sturgeon, shrimp, lobster, crayfish, crabs, oysters, clams, scallops, mussels and abalone. Aquaculture also includes alligator and crocodile breeding, as tourist attractions and for their skins, which could prove a lucrative activity.

The temperatures required for aquatic species are generally in the 20–30 °C range. The size of the installation will depend on the temperature of the geothermal source, the temperature required in the fish ponds and the heat losses from the latter.

Geothermal energy has been used successfully to provide the heat needed to grow Spirulina throughout the year in temperate countries. This is a single-celled, spiral-shaped, blue-green micro-alga, in the past called "super-food" because of its nutrient density, although at the moment it is being marketed as a nutritional food supplement. Spirulina is being farmed in a number of tropical and sub-tropical countries, in lakes or artificial basins, where conditions are ideal for its fast and widespread growth (a hot, alkaline environment rich in CO_2).

Industrial applications, including process heating, evaporation, drying, distillation, sterilization, washing, de-icing, and salt extraction are reported so far in 14 countries, where the installations tend to be large and energy consumption high. Examples of application also include bottling of water and carbonated drinks, paper and vehicle parts production, oil recovery, food processing and milk pasteurization, leather industry, chemical extraction, CO_2 extraction, laundry use, pulp and paper processing, and borate and boric acid production. The best example in the world of industrial applications are the diatomaceous drying in Námafjall (Iceland), pulp and paper processing at Kawerau (New Zealand), and the 50 years old industry of milk products Medo-Bel Creamery of Klamath Falls (Oregon). Desalination using geothermal heat has been proposed in many places, mainly volcanic island, for solving a typical problem of lack of drinking water in tourist season in remote island. It has never proved to be economical and the proposals never reached the stage of demonstration but, with the help of heat pumps, the plants might see a growing application in the future. More and more wineries are adopting geothermal systems for temperature control of the various phases of wine production, especially now that heat pumps technology contributes also to humidity control.

The elements to consider in the *cost estimate* of a geothermal system for heat application are numerous and often more complicated than for other forms of energy. Plant installation and operating cost must be evaluated carefully before a geothermal project is launched, and vary from case to case. The overall cost depends on the resource (amount of heat, fluid discharge rate, depth to drill), the cost of installation and operation, and financial conditions including available tariffs and incentives, the price of other available energy sources, the plant efficiency, the load time. District space heating levelized costs range from US$ 45/MWh_t to US$ 85/MWh_t. Costs of heating greenhouses vary between US$ 40/MWh_t and US$ 50/MWh_t [4].

3. – Power production

Electricity generation mainly needs steam to rotate a turbine that activates a generator, which produces electricity. Most power plants still use fossil fuels to boil water for steam, whereas geothermal power plants use steam produced from or heated by underground hot fluids. Three main types of technology are available: dry steam, flash steam and binary cycle, depending on the characteristics of the geothermal fluid.

In dry steam plants the steam is piped directly from underground wells to the power plant, where it is directed into a turbine/generator unit. To use this plant, geothermal fluids must be in vapor state when they reach the surface. In a few cases (The Geysers in California, Larderello in Italy, Matsukawa in Japan, Kamojang and Darajat in Indonesia) the geothermal fluid is at vapor state already in the reservoir. These systems, named vapor-dominated systems, are characterized by very high temperature (exceeding 300 °C) and high production but are very localized. The most common case is that of water-

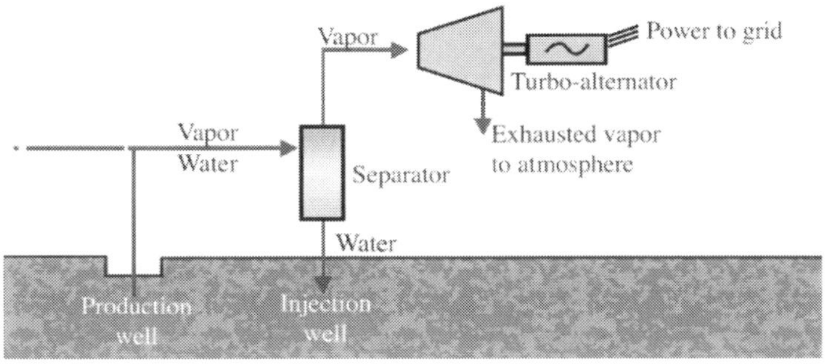

Fig. 11. – Simplified flow diagram for direct steam, atmospheric exhaust geothermal power plant.

dominated systems, having a mixture of liquid and vapor in the reservoir. When this mixture is reached by a well it flows upward by its own pressure. The flow results in a pressure drop. For very hot fluids, usually above 250 °C, the fluid pressure at the surface results much lower than the saturation pressure corresponding to the fluid temperature, and the mixture of water and steam is wholly vaporized before reaching the surface. In this case a dry steam plant can be used.

Steam turbines are available with either atmospheric (back-pressure) or condensing exhausts. Atmospheric exhaust units (fig. 11) are simple and cheap and quickly installed, but steam consumption (from the same inlet pressure) per kilowatt-hour produced is almost double that of a condensing unit. The steam, direct from dry steam wells or, after separation, from wet wells, is passed through a turbine and exhausted to the atmosphere. The atmospheric exhaust turbines are used as pilot plants, stand-by plants, in the case of small supplies from isolated wells, and for generating electricity from test wells during field development.

In the condensing units with condenser at the outlet of the turbine, the steam in the turbine expands to the (low) pressure value of the condenser (fig. 12). In the condenser the steam is cooled by spraying cooling water. The cooling water and condensed

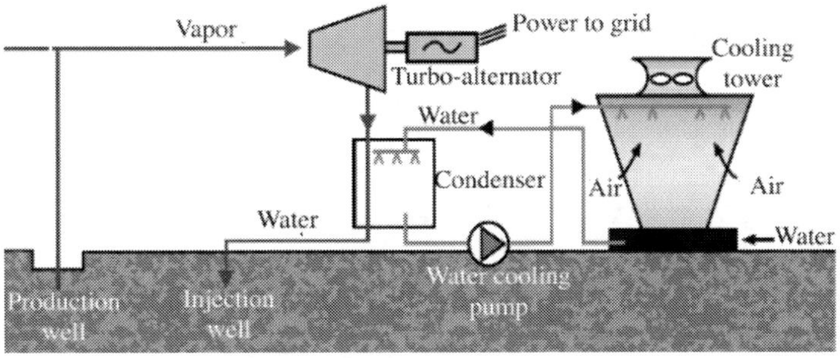

Fig. 12. – Simplified flow diagram for direct steam, condensing geothermal power plant.

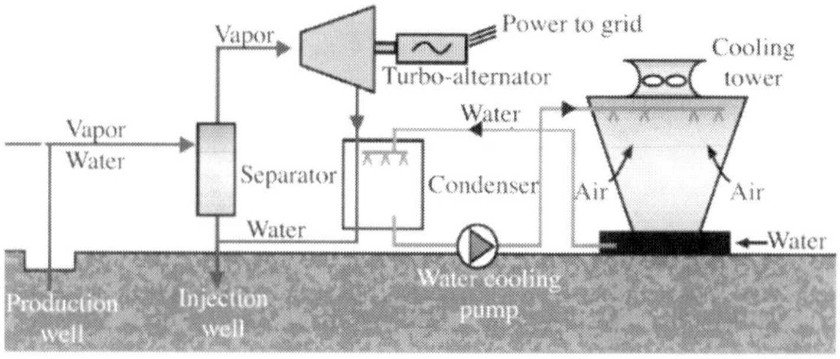

Fig. 13. – Simplified flow diagram for single flash geothermal power plant.

steam are mixed and pumped directly to the cooling tower for recirculation. Having more auxiliary equipment, the condensing units are more complex than the atmospheric exhaust units and the bigger sizes can take twice as long to construct and install. The resulting power capacity of the condensing units is, however, much higher than that of the atmospheric exhaust units. Condensing plants have an average size of 45 MW_e. Units of 40–60 MW_e capacity are very common, but plants of 110 MW_e have also been constructed and installed.

Flash steam turbines are the most common and use very hot geothermal fluids, with temperatures higher than 180 °C. When the reservoir is reached by wells, the mixture of water and steam of liquid-dominated systems flows up through wells in the ground under its own pressure. As it flows upward, the pressure decreases and some of the hot water boils into steam. This change of state for pressure drop, called flash, gives the name to the technology. The original and flashed steam is then separated from the water and used to power a turbine/generator. The liquid phase is separated by the steam and piped to be re-injected back in the subsurface. From this point on the equipment is the same of a dry steam plant, and this plant is called single-flash plant (fig. 13). Depending on the thermodynamic condition of the fluids, it could be economic to obtain a second (sometime even a third) flash from the separated fluid, producing more steam from its boiling. This double-flash steam power type of plant differ from the previous ones for the equipment required for the secondary flash. Flash steam plants have an average size of 30 MW_e.

Generating electricity from low-to-medium temperature geothermal fluids and from the waste hot waters coming from the separators in water-dominated geothermal fields has made considerable progress since the first improvements were made in binary fluid technology. The *binary cycle plants* use the heat from the hot water to boil a working fluid, usually an organic fluid with a low boiling point and high vapor pressure at low temperatures when compared to steam. The working fluid is vaporized in a heat exchanger; the vapor produced drives a normal axial flow turbine, is then cooled and condensed, and the cycle begins again (fig. 14). The geothermal water is then injected back into the ground. The water and the working fluid circulate separated during the

DOI: 10.1051/epjconf/201714800012

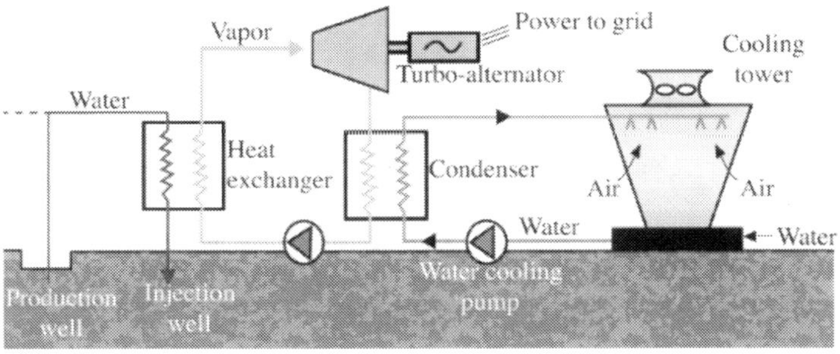

Fig. 14. – Simplified flow diagram for ORC binary geothermal power plant.

whole process, and the working fluid circulates in a closed loop, so there are little or no air emissions and the geothermal water is totally re-injected.

The two different categories of binary cycles commonly available, Organic Rankine Cycles (ORC) and Kalina cycles, mostly differentiate as far as the working fluid is concerned: in ORC a pure working fluid, (or seldom an azeotropic mixture) is utilized, while in Kalina cycle and Kalina derived cycles, a mixture of water and ammonia is selected. By selecting suitable secondary fluids installed binary systems utilize geothermal fluids in the temperature range 73 °C (at Chena Hot Springs, Alaska) and 180 °C. The upper limit depends on the thermal stability of the organic binary fluid, and the lower limit on technical-economic factors: below a certain temperature and depending on the air temperature at the condenser the size of the required heat exchangers would render the project uneconomical. Beside production from low-to-medium temperature geothermal fluids and waste fluids, binary systems can also be used to avoid flashing of the geothermal fluids (for example, to prevent well sealing). In this case, down-hole pumps can be used to keep the fluids in a pressurised liquid state, and the energy can be extracted from the circulating fluid by means of binary units.

Binary plants are usually constructed in small modular units of a few hundred kW_e to a few MW_e capacity. The total size of the plant has little effect on the specific cost, as a series of standard modular units is joined together to create power-plants up to a few tens of megawatts. Their cost depends on a number of factors, but particularly on the temperature of the geothermal fluid produced, which influences the size of the turbine, heat exchangers and cooling system.

It is also possible to cascade a steam unit with a binary cycle plant, which uses the condensed steam. These combined units are under development for further optimization. *Cogeneration* scheme, *i.e.* the generation of heat (usually for district heating) and power, is also feasible, as well as *hybrid plants* using the geothermal source in addition to another thermal source (*e.g.* solar or biomass or wastes).

Any leftover water and condensed steam are injected back into the reservoir, for a sustainable use of the resource. A geothermal plant is characterized by a multitude of pipes connecting production wells to the power station for energy conversion, and pipes distributing condensed water for re-injection.

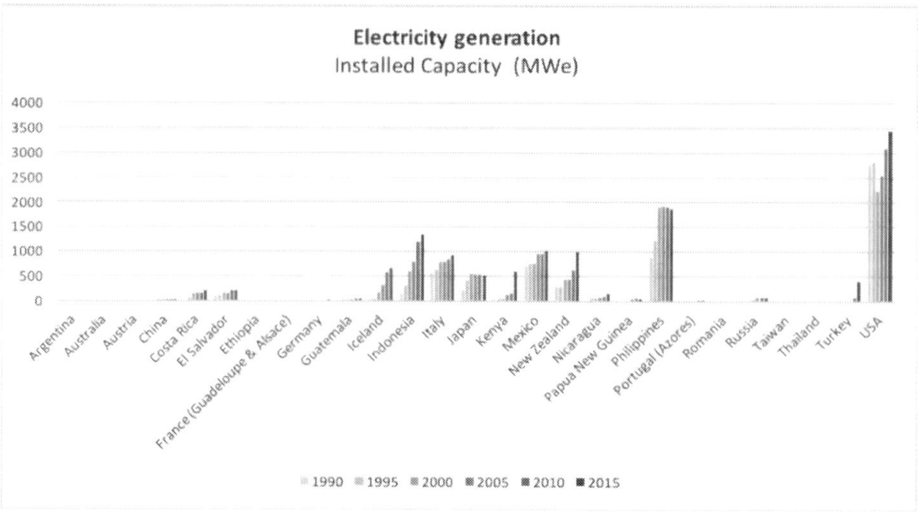

Fig. 15. – Power production installed capacity trend in 27 countries.

Nowadays, power production in the world have a total capacity of 12.6 GW [5]. Single flash power plants are the most used technology for geothermal power, composing 41% ($\sim$ 5.079 MW_e) of Installed Capacity globally. Dry Steam follows at 22% ($\sim$ 2.863 MW) and double flash ranks third at 21% ($\sim$ 2.544 MW) of global installed capacity. Lastly, binary is 12% ($\sim$ 1.790 MW) of global installed capacity. The last 3% of power plants are triple flash, back pressure, flash/binary hybrid, EGS (described in the Present and Future chapter), or some other geothermal technology.

Geothermal power production has a much shorter history than direct uses. Commercial power production commenced in Italy at the beginning of the 20th century, when Prince Ginori Conti, heir of the industrial facilities of Larderello (Tuscany, Italy) for the production of boric acid, had the idea of using the hot fluids to generate the power required by the factory. Power production in Italy increased for a while and ceased during the II World War, when the facilities were destroyed by bombs. After the war, power production steadily increased in Italy, soon accompanied by a quick growing of production in California, New Zealand and Japan. Nowadays, 27 countries produce power from geothermal resources (fig. 15). USA has the largest geothermal power production area (3.45 GW), mainly from The Geysers, California, followed by Philippines (1.87 GW), Indonesia (1.34 GW), Mexico (1.017 GW), New Zealand (1.005 GW) (fig. 16) [5].

Capital cost of geothermal power production plants is high, as for any other use of geothermal energy. However, since the geothermal energy is constantly provided, geothermal has a base-load production, and the Capacity Factor (CF, *i.e.* the actual produced energy with respect to the full capacity energy) is much higher than for other renewable energies, resulting in a Leveled Cost of Energy comparable to other energy sources.

Power production upfront cost can vary widely, from US\$ 2.8 million to US\$ 5.5 million per MW_e (estimated for a reference plant of 50 MW_e installed capacity), depending

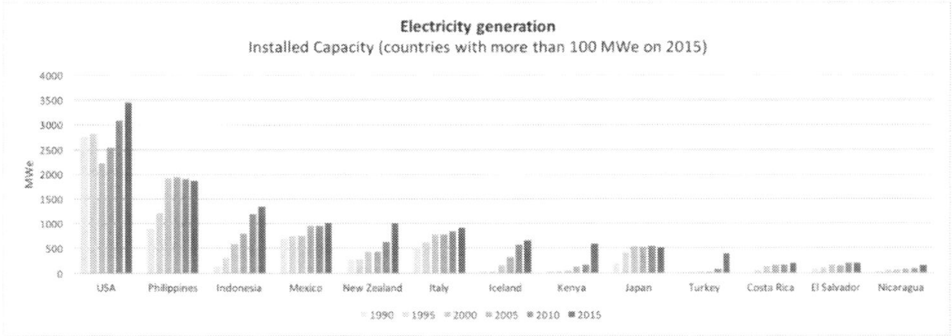

Fig. 16. – Power production installed capacity trend in the 13 countries with capacity larger than 100 MW on 2015.

on factors such as the geology of a country or region, quality of the resource (*e.g.*, temperature, flow rate, chemistry), and the infrastructure in place [4]. However, geothermal power can be competitive and complement other sources of generation thanks to high capacity factors, long plant lifetimes, and the absence of recurring fuel costs. Levelized costs of energy from hydrothermal resources are usually found to be between US$ 50 and 80 per MW_e. Binary plants have higher upper limits: levelized costs for new greenfield plants can be as high as US$ 120 per MW_e in the United States and US$ 200 per MW_e in Europe, for small plants and lower temperature resources [4].

The investment cost breakdown of utility scale geothermal power development (based on data from Iceland) shows that 35% of the total cost is related to power plant construction, and 34% to drilling. The remaining 31% is made of steam gathering system (13%), interconnection (6%), early development (5%), miscellaneous (5%) and infrastructure (2%) costs [6].

4. – Geothermal project development

To realize a geothermal plant, utilizing either shallow and deep systems, a number of activities are necessary and a geothermal project is made of essentially standard stages (fig. 17).

First of all, the geothermal resource is assessed through exploration, with the aim of defining the amount of heat and physicochemical characteristics of the underground. The latter are required for the correct planning of the later activities. In parallel, especially for heat use, it is necessary to define the amount and rate of required thermal energy. After the assessment stage, the boreholes (for shallow systems) and wells are drilled and the resource is accessed and verified. The technology to be used in the plant is defined on the base of the information retrieved during the assessment and refined in the drilling stages. An environmental impact analysis is also made, as required by law almost everywhere. Eventually, facilities are built and the plant is operated and managed.

The *exploration program* is usually developed on a step-by-step basis: *reconnaissance*, *pre-feasibility* and *feasibility*. Before drawing up a complete geothermal exploration pro-

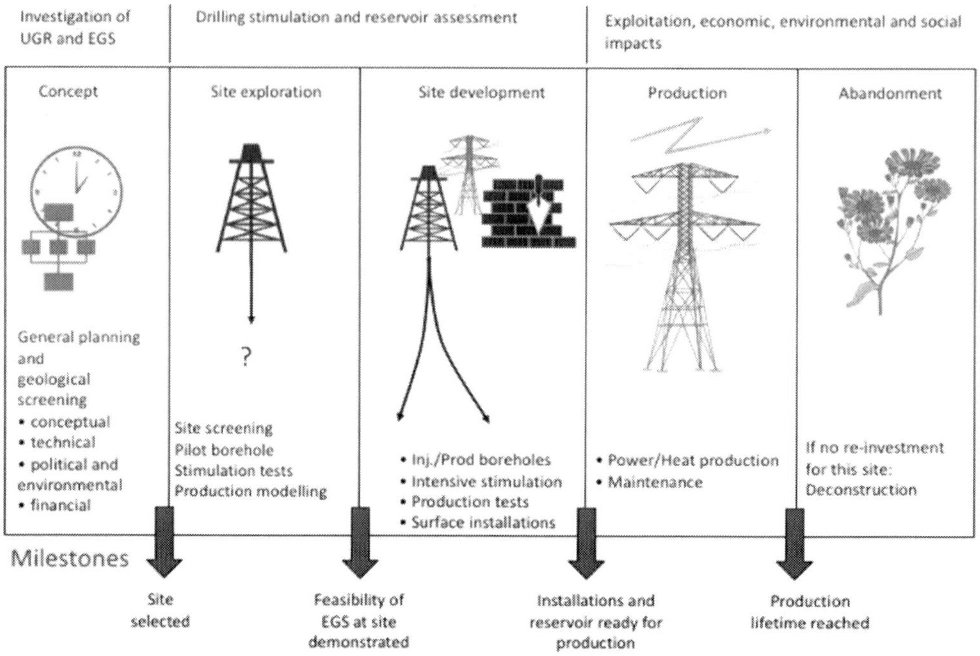

Fig. 17. – Development phases of a geothermal project.

gram all existing geological, hydrogeological, geophysical and geochemical data of the study area and adjacent areas must be collected. For the assessment of deep geothermal systems this *pre-feasibility* analysis refers to a broad area in order to eliminate the less interesting areas and concentrate on the most promising ones (*reconnaissance*) and *pre-feasibility* is carried out on the chosen areas of few square kilometers. This information frequently plays an important role in defining the objectives of the geothermal exploration program and could lead to a significant reduction in costs. When all information is retrieved and the geothermal systems is roughly determined, new data are acquired to complete the picture and decide where to drill boreholes or wells (*feasibility*). Usually, the exploration stage is completed with the first, exploratory well drilled and a direct characterization of underground condition.

The size and budget of the exploration should be proportional to its objectives, to the importance of the resources we expect to find, and to the planned forms of utilization. The exploration schedule should be flexible, as it is often reassessed as the results come in from the various surveys; similarly the geological-geothermal model should be progressively updated and improved. These periodic re-assessments eliminate any operations that are no longer necessary and insert others, according to the results attained, in order to avoid unnecessary costs, and also to decrease the risk of error or failure in assessing and retrieving the resource. Conversely, by decreasing the risk of error the overall cost increases. The economic success of a geothermal exploration hinges on finding the proper balance between the two.

DOI: 10.1051/epjconf/201714800012

Drilling represents one of the main costs of a geothermal plant, and the number of boreholes or wells should be kept at a minimum, just necessary to satisfy the resource required by the plant to produce the demanded energy. Drilling and completion of wells are very critical operations, to be conducted with an eye on safety of the well and of the surrounding environment, and on speed, in order to decrease both operational cost and disturbance time to the neighbors. Shallow boreholes are drilled with the same technologies for fresh water wells, with similar safety precautions. The most peculiar part of shallow geothermal drilling is related to completion of geoexchangers, to secure the conductive exchange.

Deep geothermal wells uses technology similar to the rotary drilling for oil and gas. Main differences are due to the different geomechanical, physical and chemical conditions. Rocks in geothermal systems span from sedimentary rocks (as in oil and gas) to metamorphic, crystalline and volcanic rocks, usually much harder to drill. The high temperature affects the circulation system and the cementing procedure. Heavy mud is often used, due to the heavy load of debris to extract during the drilling. When possible, also air is used since it makes the drilling much faster and cheaper, but is used with care in formation carrying excessive water or prone to collapse. Chemical condition must be checked, especially at high temperature, to avoid productive fracture sealing, mud drying, damage to the casing or the drilling bits due to aggressive components. Blowout preventers are often used, especially in volcanic and magmatic condition where geothermal fluid may be rich of gases.

Directional drilling is used, although sub-vertical wells are common. Sometimes it may be attractive or even necessary to drill boreholes from the same platform and deviate them in different directions. Casing is usually necessary, but in hard rocks, where the risk of collapse is minimum, can be avoided. Immediately after well completion, the well is tested to obtain information regarding physicochemical condition of fluids and rocks at depth. Depending on well condition, geological and physical analyses (logs) are carried out in the well, rock and fluid samples are taken for further analyses in the labs. The most important information are temperature, depth, condition of fluid productive fractures, pressures at various depth, types of rocks and permeability, porosity, fluid content and its chemical composition. This information is used for refining the interpretation of deep geological structures, their thermal and hydraulic condition. Reservoir modeling is then carried out, to estimate potential fluid flow from the well, energy potential, and predict well performance (pressure, temperature, enthalpy of fluid).

Where wells result productive, the geothermal system is exploited. The reservoir models, after validation against field data, are then inserted in the overall reservoir simulation used for designing the *geothermal field development*. Well location and gathering system design must be carefully planned. Some example of consideration: wells are preferably situated higher in the landscape than the separator station and power station; if possible, the power station should be situated a little lower than the separator station, to facilitate natural fluid flow; to join multiple wells into one steam gathering system requires decision of optimum separator pressure. Optimization requires balancing of technical and economical aspects: for example, the optimum separator pressure can be calculated from a thermodynamic standpoint maximizing the generated power output, but cost considerations (capital cost of the power plant increase at decreasing steam pressure) and other constraints also dictate pressure choice. Field development comprise, after reservoir modeling, further drilling and the construction of surface plants and equipment.

5. – Environmental impact

Exploitation of geothermal energy has an impact on the environment, but it is one of the least polluting forms of energy. The degree to which geothermal exploitation affects the environment looks proportional to the scale of its exploitation and the depth of resource, and shallow geothermal systems can easily avoid to produce environmental impact by adapting the most elementary precautions. The most critical potential risk is thermal pollution, when the boreholes are incorrectly localized or the rate of extraction is too high. The first effect of wrong design is, however, a low efficiency of the plant, a problem that usually finds a technical solution. Electricity generation in binary cycle plants will affect the environment in the same, limited way as direct heat uses. The effects are potentially greater in the case of conventional back-pressure or condensing power plants, especially as regards air quality, but can be kept within acceptable limits. The following aspects essentially refer to deep geothermal system deployment.

Any modification to our environment must be evaluated carefully, in deference to the relevant laws and regulations (which in some countries are very severe), but also because an apparently insignificant modification could trigger a chain of events whose impact is difficult to fully assess beforehand. For example, a mere 2–3 °C increase in the temperature of a body of water as a result of discharging the waste water from a utilization plant could damage its ecosystem.

The most perceptible effect on the environment is that of *drilling*, whether the boreholes are shallow ones for geoexchangers or deep producing wells. Installation of a drilling rig and all the accessory equipment entails the construction of a drilling pad and access roads if they are not available. These operations will modify the surface morphology of the area and could damage local plants and wildlife. In case of blowouts of fluids from the wells, fluids can pollute surface water and air. This eventuality is nowadays well solved by the essentially constant presence of blowout preventers where high temperatures and pressures, and aggressive chemicals are anticipated. During flow tests undesirable gases may be discharged into the atmosphere, but only for a very limited time. Noise can be annoying, and drilling technology tries hard to keep noise at the lowest possible level with silencers. The best solution, indeed, is to avoid to stay too close to urban areas and to keep the drilling time as quick as possible. The impact on the environment caused by drilling is temporary, and mostly ends once drilling is completed.

Installation of the pipelines that will transport the geothermal fluids, and construction of the *utilization plants*, may also affect animal and plant life and the surface morphology. The landscape is modified, although in some areas the network of pipelines criss-crossing the countryside and the power plant cooling towers have become an integral part of the panorama and are indeed a famous tourist attraction in many geothermal areas. Power plant can be built in a fancy design, as in some Italian areas, or partially hidden by vegetation (usually outside the facility to ease maintenance).

Plant operation of deep, high-temperature geothermal resources, especially in volcanic and magmatic areas, may produce environmental problems. Some geothermal fluids, such as those utilized for district-heating in Iceland, are freshwaters, but this is very rare and deep geothermal fluids (steam or hot water) often contain *gases* such as carbon dioxide (CO_2), hydrogen sulphide (H_2S), ammonia (NH_3), methane (CH_4), and trace amounts of other gases, as well as *dissolved chemicals* whose concentrations usually increase with temperature. For example, sodium chloride (NaCl), boron (B), arsenic (As) and mercury (Hg) are a source of pollution if discharged into the environment. The waste waters from geothermal plants also have a higher temperature than the environment and therefore constitute a potential thermal pollutant. Geothermal fluids, however, may be treated and chemicals separated. Various processes can be adopted to reduce gas emissions, *e.g.* for hydrogen sulphide. Carbon dioxide is also present in the fluids used in the geothermal power plants, although much less CO_2 is discharged from these plants than from fossil-fuelled power stations.

Discharge of waste waters is also a potential source of chemical pollution, although the low-to-moderate temperature geothermal fluids used in most direct-use applications generally contain low levels of chemicals and the discharge of spent geothermal fluids is seldom a major problem. Spent geothermal fluids with high concentrations of chemicals such as boron, fluoride or arsenic of deep, high-temperature fluids should be treated and re-injected into the reservoir, or elements extracted for mineral use. To avoid discharge into surface waters after cooling, the waters can be cooled in special storage ponds or tanks to avoid modifying the ecosystem of natural bodies of waters (rivers, lakes and even the sea).

Extraction of large quantities of fluids from geothermal reservoirs reduce pore pressure and the land surface may tend to *subside*, *i.e.* to gradually sink. This is an irreversible phenomenon, a slow process distributed over vast areas. Over a number of years the lowering of the land surface could reach detectable levels, in some cases of the order of a few tens of centimeters and even meters, and should be monitored systematically. This effect is nowadays famous, after Wairakei in New Zealand registered a 4.5 m localized subsidence between 1964 and 1974: water-dominated systems such as Wairakei subside more than vapor-dominated systems. Subsidence occurrence is controlled by topographic monitoring, so far accomplished by high-resolution GPS remote control. Subsidence is prevented or reduced by re-injection of spent fluids, which also sustain the production.

The withdrawal and/or re-injection of geothermal fluids may trigger or increase the frequency of *seismic events* in tectonically active areas, such as those close to plate boundaries where the high temperature geothermal areas are located. In these areas the injection of fluids tends to release the accumulated tectonic stress. These small events, mostly detected by means of instrumentation, show an increase in number. Correlation studies in high temperature geothermal areas have shown, however, that the magnitude of events is not increased with respect to the maximum magnitude registered in historical time. Exploitation of hydrothermal resources is unlikely to trigger major seismic events, and so far has never been known to do so.

 DOI: 10.1051/epjconf/201714800012

The *noise* associated with operating power generation geothermal plants is not negligible in the occasional vent discharge, although normally acceptable. At the power plant the main noise pollution comes from the cooling tower fans, the steam ejector, and the turbine "hum". The noise generated in direct heat applications is usually negligible.

6. – Present and future

The thermal energy present in the underground is enormous. If exploited correctly, geothermal energy could certainly assume an important role in the energy balance and small-scale geothermal resources are capable of improving energy efficiency of many applications. Today the global installed capacity of geothermal energy amounts to about 60 GW worldwide with shares of 18%, 26%, and 56% for power generation, direct use, and ground source heat pumps (GSHP), respectively. Lead markets for geothermal energy are in America, Europe and Asia [7].

The primary obstacles to geothermal deployment are the high investment cost and the risk of drilling without reaching commercially viable geothermal resources. Lowering drilling costs is a key issue for lowering up-front costs. R&D efforts are also focussing on reducing heat exchanger costs, since design and layout of heat exchanger heavily influence the investment and operation costs of geothermal plants. Optimization of heat exchangers play a central role in geothermal binary power plants but also for direct use and GSHP applications. Regarding the risk of not finding the enough fluids to run a geothermal plant, R&D in exploration technology and implementation of a risk insurance fund are under consideration.

The development of *Unconventional Geothermal Resources(UGR)*, which requires technological development to become economically productive, would increase enormously the share of utilized geothermal energy. Among UGR, research projects have focused so far to two kind of resources: Coproduced and Enhanced Geothermal Systems.

Enhanced Geothermal Systems (EGS) comprehend the development of geothermal systems where the natural flow capacity of the system is not sufficient to support adequate power production but where artificial fracturing of the system by chemical and/or hydraulic stimulation can allow production at a commercial level. The reservoirs are created to produce energy from geothermal resources that are otherwise not economical due to lack of water and/or permeability. EGS technology requires validation and deployment, cost reduction, and performance improvement. While achieving cost-competitive electricity generation from EGS is a long-term goal, in the near-term R&D and demonstration projects will move industry along the learning curve toward technological readiness. Once technical and economic challenges for EGS are overcome, or other methods of exploiting hot rock resources become available (*e.g.* without fracturing the underground bedrock), geothermal deployment could be pursued wherever rock temperatures and other underground properties allow the economic sale of energy. This would mean that advanced geothermal systems could be deployed where demand for electricity and heat exist.

In *coproduced resources* power may be generated by the hot fluid of oil, gas, and other material harvesting processes by-products. While the quality of the resource depends on water volume and temperature, these technologies have the potential to extend the economic life of oil and gas fields. The research is primarily carried out in USA, where hundreds of thousands of oil and gas wells produce hot water concurrent with oil and gas production. It is estimated that the water produced annually by oil and gas fields could generate up to 3 GW_e of clean, base-load power using binary geothermal units [8]. Other uses, such as district heating and cooling, may be considered where oil and gas fields are close to urban settings. These technologies offer an opportunity to expand fossil fuels development into the renewable energy sector. Coproduced resources include also mineral recovery from geothermal fluids.

Cogeneration of heat and power (Combined Heat and Power, CHP) from deep aquifers and *hybrid* technology integrating geothermal to other green energy sources are important targets for innovative technology. The sale of heat from CHP development (*e.g.* for district heating) increases the economic viability of lower-temperature resources. It will be also important to further optimize *cascade use* of geothermal energy.

Estimated global geothermal electrical power installed capacity by 2050 amounts to 200 GW_e, including 100 GW_e hydrothermal electricity capacity and 100 GW_e from EGS [4]. The reference scenario assumes that hot rock technology will become commercially viable soon after 2030. This power production is expected to avoid around 760 megatonnes (Mt) of CO_2 emissions per year worldwide. Most of this increase is expected to happen in Pacific Asia, mainly Indonesia; the East-African Rift Valley; Central and South America; as well as in the United States, Japan, New Zealand, and Iceland.

The estimated global sum of annual geothermal heat use by 2050 amounts to 5.8 EJ (about 1 600 TWh thermal energy), removing contributions from ground source heat pumps and assuming full use of the potential of CHP generation via EGS technology [4]. Under the assumed condition, the utilization of heat from deep rock formations should theoretically become possible wherever rock temperatures and the properties of the underground allow the economic sale of energy. The largest potential for geothermal heat can be found in regions with high heat demand: Europe, China and North America.

The International Energy Association on 2011 listed the key conditions for the full development of geothermal energy by 2050:

- A holistic policy framework is needed that addresses technical barriers relating to resource assessment, accessing and engineering the resource, geothermal heat use and advanced geothermal technologies. Moreover, such a holistic framework needs to address barriers relating to economics, regulations, market facilitation and RD&D support.
- Policy makers, local authorities and utilities need to be more aware of the full range of geothermal resources available and of their possible applications in order to develop consistent policies accordingly. This is particularly true for geothermal heat, which can be used at varying temperatures for a wide variety of tasks.
- Important R&D priorities for geothermal energy include accelerating resource assessment, development of more competitive drilling technology and improving EGS technology as well as managing health, safety and environmental (HSE) concerns.

– Advanced technologies for offshore, geo-pressured and super-critical (or even magma) resources could unlock a huge additional resource base. Where reasonable, co-produced hot water from oil and gas wells can be turned into an economic asset.

REFERENCES

[1] Barbier E., *Renew. Sustain. Energy Rev.*, **6** (2002) 3.
[2] Lund J. W. and Boyd T. L., "Direct Utilization of Geothermal Energy 2015 Worldwide Review", in *Proceedings World Geothermal Congress 2015, Melbourne, Australia* (2015).
[3] European Geothermal Energy Council, *Market report 2013/2014*, December 2013.
[4] International Energy Agency, *Technology Roadmap: Geothermal Heat and Power* (OECD/IEA, Paris) 2011.
[5] Bertani R., "Geothermal power generation in the World 2010–2014 update report", *Geothermics*, **60** (2016) 31.
[6] Gehringer M. and Loksha V., *Geothermal Handbook: Planning and Financing Power Generation*, ESMAP technical report no. 002/12 (World Bank, Washington, DC) 2012.
[7] Sigfusson B. and Uihlein A., *2014 JRC Geothermal Energy Status Report* (Publications Office of the European Union) 2015, ISBN: 978-92-79-44614-6, ISSN: 1831-9424.
[8] Matek B., *2013 Geothermal Power: International Market Overview* (Geothermal Energy Association) 2013.

EPJ Web of Conferences **148**, 00013 (2017) DOI: 10.1051/epjconf/201714800013
LNES 2016

Fuel cells for electrochemical energy conversion

Ryan P. O'Hayre

Metallurgical and Materials Engineering, Colorado School of Mines
1500 Illinois St, Golden, CO 80401, USA

Summary. — This short article provides an overview of fuel cell science and technology. This article is intended to act as a "primer" on fuel cells that one can use to begin a deeper investigation into this fascinating and promising technology. You will learn what fuel cell are, how they work, and what significant advantages and disadvantages they present.

1. – What is a fuel cell?

A fuel cell is essentially a compact electrochemical "factory" that takes fuel as input and produces electricity as output (see fig. 1). Like a factory, a fuel cell will continue to churn out product (electricity) as long as raw material (fuel) is supplied. This is the key difference between a fuel cell and a battery. While both rely on electrochemistry to work their magic, a fuel cell is not consumed when it produces electricity. It is really a factory, a *shell*, which transforms the chemical energy stored in a fuel into electrical energy.

Viewed this way, combustion engines are also "chemical factories". Combustion engines also take the chemical energy stored in a fuel and transform it into useful mechanical or electrical energy. So what is the difference between a combustion engine and a fuel cell?

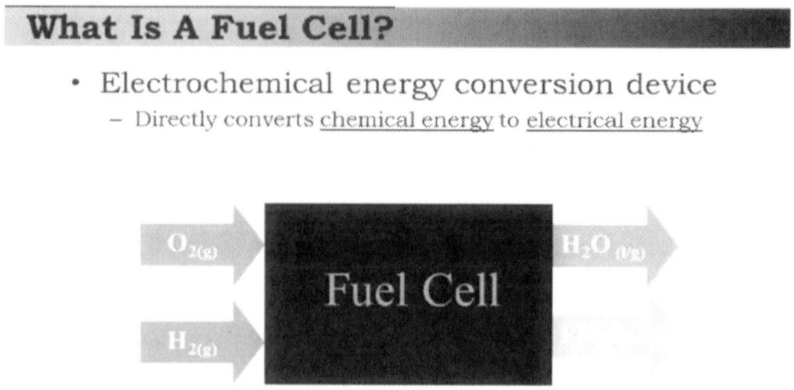

Fig. 1. – General concept of a (H_2–O_2) fuel cell.

In a conventional combustion engine, fuel is burned, releasing heat. Consider the simplest example, the combustion of hydrogen:

$$H_2 + \frac{1}{2}O_2 \rightarrow H_2O. \tag{1}$$

On the molecular scale, collisions between hydrogen molecules and oxygen molecules result in a reaction. The hydrogen molecules are oxidized, producing water and releasing heat. Specifically, at the atomic scale, in a matter of picoseconds, hydrogen-hydrogen bonds and oxygen-oxygen bonds are broken, while hydrogen-oxygen bonds are formed. These bonds are broken and formed by the transfer of electrons between the molecules. The energy of the product water bonding configuration is lower than the bonding configurations of the initial hydrogen and oxygen gases. This energy difference is released as heat. Although the energy difference between the initial and final states occurs by a reconfiguration of electrons as they move from one bonding state to another, this energy is recoverable only as heat because the bonding reconfiguration occurs in picoseconds at an intimate, subatomic scale. To produce electricity, this heat energy must be converted into mechanical energy, and then the mechanical energy must be converted into electrical energy. Going through all these steps is potentially complex and inefficient.

Consider an alternative solution: to produce electricity directly from the chemical reaction by somehow harnessing the electrons as they move from high-energy reactant bonds to low-energy product bonds. In fact, this is exactly what a fuel cell does. But the question is how do we harness electrons that reconfigure in picoseconds at subatomic length scales? The answer is to spatially separate the hydrogen and oxygen reactants so that the electron transfer necessary to complete the bonding reconfiguration occurs over a greatly extended length scale. Then, as the electrons move from the fuel species to the oxidant species, they can be harnessed as an electrical current.

EPJ Web of Conferences **148**, 00013 (2017) DOI: 10.1051/epjconf/201714800013
LNES 2016

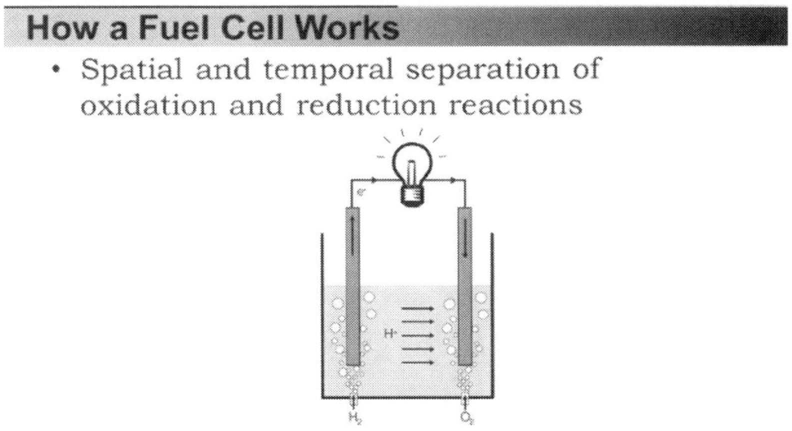

Fig. 2. – A simple fuel cell.

2. – A simple fuel cell

In a fuel cell, the hydrogen *combustion* reaction is split into two *electrochemical* half reactions:

$$H_2 \rightarrow 2H^+ + 2e^-, \tag{2}$$

$$\frac{1}{2}O_2 + 2H^+ + 2e^- \rightarrow H_2O. \tag{3}$$

By spatially separating these reactions, the electrons transferred from the fuel are forced to flow through an external circuit (thus constituting an electric current) and do useful work before they can complete the reaction.

Spatial separation is accomplished by employing an electrolyte. An electrolyte is a material that allows ions (charged atoms) to flow but not electrons. At a minimum, a fuel cell must possess two electrodes, where the two electrochemical half reactions occur, separated by an electrolyte.

Figure 2 shows an example of an extremely simple H_2–O_2 fuel cell. This fuel cell consists of two platinum electrodes dipped into sulfuric acid (an aqueous acid electrolyte). Hydrogen gas, bubbled across the left electrode, is split into protons (H^+) and electrons following eq. (2). The protons can flow through the electrolyte (the sulfuric acid is like a "sea" of H^+), but the electrons cannot. Instead, the electrons flow from left to right through a piece of wire that connects the two platinum electrodes. Note that the resulting current, as it is traditionally defined, is in the opposite direction. When the electrons reach the right electrode, they recombine with protons and bubbling oxygen gas to produce water following eq. (3). If a load (*e.g.*, a light bulb) is introduced along the path of the electrons, the flowing electrons will provide power to the load, causing the light bulb to glow. Our fuel cell is producing electricity! The first fuel cell, invented by William Grove in 1839, probably looked a lot like the one discussed here.

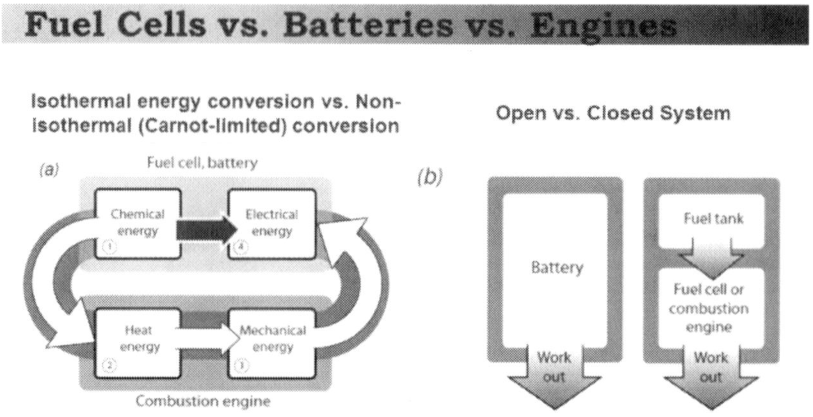

Fig. 3. – Schematic comparison of fuel cells, batteries, and combustion engines. (a) Fuel cells and batteries produce electricity directly from chemical energy. In contrast, combustion engines first convert chemical energy into heat, then mechanical energy, and finally electricity (alternatively, the mechanical energy can sometimes be used directly). (b) In batteries, power and capacity are typically intertwined —the battery is both the energy storage and the energy conversion device. In contrast, fuel cells and combustion engines allow independent scaling between power (determined by the fuel cell or engine size) and capacity (determined by the fuel tank size).

3. – Fuel cell advantages

Because fuel cells are "factories" that produce electricity as long as they are supplied with fuel, they share some characteristics in common with combustion engines. Because fuel cells are electrochemical energy conversion devices that rely on electrochemistry to work their magic, they share some characteristics in common with primary batteries. In fact, fuel cells combine many of the advantages of both engines and batteries.

Since fuel cells produce electricity directly from chemical energy, they are often far more efficient than combustion engines. Fuel cells can be all solid state and mechanically ideal, meaning no moving parts. This yields the potential for highly reliable and long-lasting systems. A lack of moving parts also means that fuel cells are silent. Also, undesirable products such as NO_x, SO_x, and particulate emissions are virtually zero.

Unlike batteries, fuel cells allow easy independent scaling between power (determined by the fuel cell size) and capacity (determined by the fuel reservoir size). In batteries, power and capacity are often convoluted. Batteries scale poorly at large sizes, whereas fuel cells scale well from the 1 W range (cell phone) to the megawatt range (power plant). Fuel cells offer potentially higher energy densities than batteries and can be quickly recharged by refueling, whereas batteries must be thrown away or plugged in for a time-consuming recharge. Figure 3 schematically illustrates the similarities and differences between fuel cells, batteries, and combustion engines.

4. – Fuel cell disadvantages

While fuel cells present intriguing advantages, they also possess some serious disadvantages. Cost represents a major barrier to fuel cell implementation. Because of prohibitive costs, fuel cell technology is currently only economically competitive in a few highly specialized applications (*e.g.*, onboard the space shuttle orbiter). Power density is another significant limitation. Power density expresses how much power a fuel cell can produce per unit volume (volumetric power density) or per unit mass (gravimetric power density). Although fuel cell power densities have improved dramatically over the past decades, further improvements are required if fuel cells are to compete in portable and automotive applications. Combustion engines and batteries generally outperform fuel cells on a volumetric power density basis; on a gravimetric power density basis, the race is much closer.

Fuel availability and storage pose further problems. Fuel cells work best on hydrogen gas, a fuel that is not widely available, has a low volumetric energy density, and is difficult to store. Alternative fuels (*e.g.*, gasoline, methanol, formic acid) are difficult to use directly and usually require reforming. These problems can reduce fuel cell performance and increase the requirements for ancillary equipment. Thus, although gasoline looks like an attractive fuel from an energy density standpoint, it is not well suited to fuel cell use.

Additional fuel cell limitations include operational temperature compatibility concerns, susceptibility to environmental poisons, and durability under start-stop cycling. These significant disadvantages will not be easy to overcome. Fuel cell adoption will be severely limited unless technological solutions can be developed to hurdle these barriers.

5. – Fuel cell types

There are five major types of fuel cells, differentiated from one another by their electrolyte:

1. Phosphoric acid fuel cell (PAFC).
2. Polymer electrolyte membrane fuel cell (PEMFC).
3. Alkaline fuel cell (AFC).
4. Molten carbonate fuel cell (MCFC).
5. Solid-oxide fuel cell (SOFC).

While all five fuel cell types are based upon the same underlying electrochemical principles, they all operate at different temperature regimens, incorporate different materials, and often differ in their fuel tolerance and performance characteristics, as shown in table I. Most of the examples in this book focus on PEMFCs or SOFCs. We will briefly contrast these two fuel cell types.

- PEMFCs employ a thin *polymer* membrane as an electrolyte (the membrane looks and feels a lot like plastic wrap). The most common PEMFC electrolyte is a

DOI: 10.1051/epjconf/201714800013

TABLE I. – *Description of major fuel cell types.*

	PEMFC	PAFC	AFC	MCFC	SOFC
Electrolyte	Polymer membrane	Liquid H_3PO_4 (immobilized)	Liquid KOH (immobilized)	Molten carbonate	Ceramic
Charge carrier	H^+	H^+	OH^-	${CO_3}^{2-}$	O^{2-}
Operating temperature	80 °C	200 °C	60–220 °C	650 °C	600–1000 °C
Catalyst	Platinum	Platinum	Platinum	Nickel	Perovskites (ceramic)
Cell components	Carbon based	Carbon based	Carbon based	Stainless based	Ceramic based
Fuel compatibility	H_2, methanol	H_2	H_2	H_2, CH_4	H_2, CH_4, CO

membrane material called Nafion^{TM}. Protons are the ionic charge carrier in a PEMFC membrane. PEMFCs are attractive for many applications because they operate at low temperature and have high power density. As we have already seen, the electrochemical half reactions in a H_2–O_2 PEMFC are

$$\begin{aligned} &H_2 \to 2H^+ + 2e^-, \\ &\frac{1}{2}O_2 + 2H^+ + 2e^- \to H_2O. \end{aligned} \tag{4}$$

– SOFCs employ a thin *ceramic* membrane as an electrolyte. Oxygen ions (O^{2-}) are the ionic charge carrier in an SOFC membrane. The most common SOFC electrolyte is an oxide material called yttria-stabilized zirconia (YSZ). In a H_2–O_2 SOFC, the electrochemical half reactions are

$$\begin{aligned} &H_2 + O^{2-} \to H_2O + 2e^-, \\ &\frac{1}{2}O_2 + 2e^- \to O^{2-}. \end{aligned} \tag{5}$$

To function properly, SOFCs must operate at high temperatures ($> 600\,°C$). They are attractive for stationary applications because they are highly efficient and fuel flexible.

Note how changing the mobile charge carrier dramatically changes the fuel cell reaction chemistry. In a PEMFC, the half reactions are mediated by the movement of protons (H^+), and water is produced at the cathode. In a SOFC, the half reactions are mediated by the motion of oxygen ions (O^{2-}), and water is produced at the anode. Note in table I how other fuel cell types use OH^- or CO_3^{2-} as ionic charge carriers. These fuel cell types will also exhibit different reaction chemistries, leading to unique advantages and disadvantages.

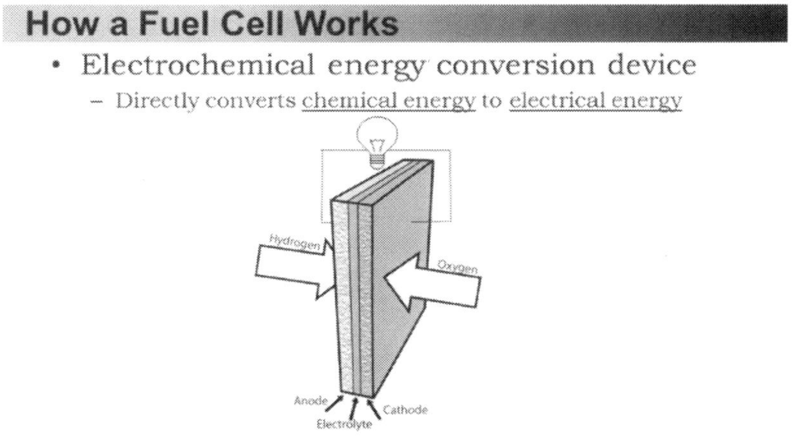

Fig. 4. – Simplified planar anode-electrolyte-cathode structure of a fuel cell.

6. – Basic fuel cell operation

The current (electricity) produced by a fuel cell scales with the size of the reaction area where the reactants, the electrode, and the electrolyte meet. In other words, doubling a fuel cell's area approximately doubles the amount of current produced.

Although this trend seems intuitive, the explanation comes from a deeper understanding of the fundamental principles involved in the electrochemical generation of electricity. As we have discussed, fuel cells produce electricity by converting a primary energy source (a fuel) into a flow of electrons. This conversion necessarily involves an energy transfer step, where the energy from the fuel source is passed along to the electrons constituting the electric current. This transfer has a finite rate and must occur at an interface or reaction surface. Thus, the amount of electricity produced scales with the amount of reaction surface area or interfacial area available for the energy transfer. Larger surface areas translate into larger currents.

To provide large reaction surfaces that maximize surface-to-volume ratios, fuel cells are usually made into thin, planar structures, as shown in fig. 4. The electrodes are highly porous to further increase the reaction surface area and ensure good gas access. One side of the planar structure is provisioned with fuel (the anode electrode), while the other side is provisioned with oxidant (the cathode electrode). A thin electrolyte layer spatially separates the fuel and oxidant electrodes and ensures that the two individual half reactions occur in isolation from one another. Compare this planar fuel cell structure with the simple fuel cell discussed earlier in fig. 2. While the two devices look quite different, noticeable similarities exist between them.

Figure 5 shows a detailed, cross-sectional view of a planar fuel cell. Using this figure as a map, we will now embark on a brief journey through the major steps involved in producing electricity in a fuel cell. Sequentially, as numbered on the drawing, these steps are as follows:

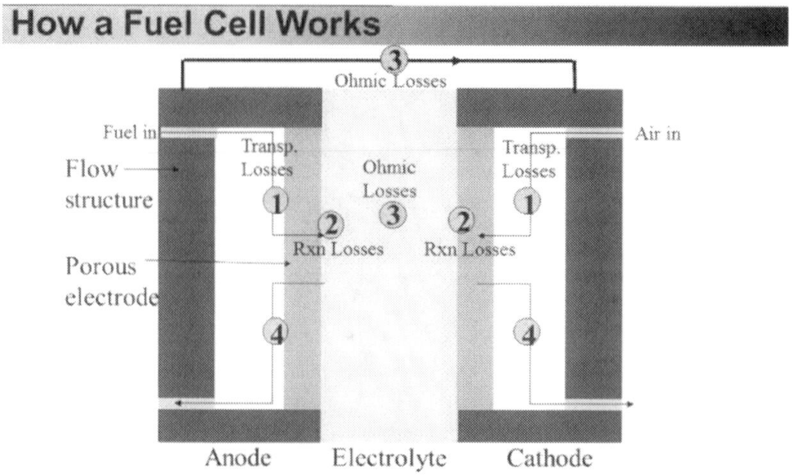

Fig. 5. – Cross-section of fuel cell illustrating major steps in electrochemical generation of electricity: (1) reactant transport, (2) electrochemical reaction, (3) ionic and electronic conduction, (4) product removal.

1. Reactant delivery (transport) into the fuel cell.
2. Electrochemical reaction.
3. Ionic conduction through the electrolyte and electron conduction through the external circuit.
4. Product removal from the fuel cell.

A few further details on each of these steps is provided below:

Step 1: *Reactant transport.* For a fuel cell to produce electricity, it must be continually supplied with fuel and oxidant. This seemingly simple task can be quite complicated. When a fuel cell is operated at high current, its demand for reactants is voracious. If the reactants are not supplied to the fuel cell quickly enough, the device will "starve." Efficient delivery of reactants is most effectively accomplished by using flow field plates in combination with porous electrode structures. Flow field plates contain many fine channels or grooves to carry the gas flow and distribute it over the surface of the fuel cell. The shape, size, and pattern of flow channels can significantly affect the performance of the fuel cell. Understanding how flow structures and porous electrode geometries influence fuel cell performance is an exercise in mass transport, diffusion, and fluid mechanics. The materials aspects of flow structures and electrodes are equally important. Components are held to stringent materials property constraints that include very specific electrical, thermal, mechanical, and corrosion requirements.

Step 2: *Electrochemical reaction.* Once the reactants are delivered to the electrodes, they must undergo electrochemical reaction. The current generated by the fuel cell is directly related to how fast the electrochemical reactions proceed. Fast electrochemical reactions result in a high current output from the fuel cell. Sluggish reactions result in low current output. Obviously, high current output is desirable. Therefore, catalysts are generally used to increase the speed and efficiency of the electrochemical reactions. Fuel cell performance critically depends on choosing the right catalyst and carefully designing the reaction zones. Often, the kinetics of the electrochemical reactions represent the single greatest limitation to fuel cell performance.

Step 3: *Ionic (and electronic) conduction.* The electrochemical reactions occurring in step 2 either produce or consume ions and electrons. Ions produced at one electrode must be consumed at the other electrode. The same holds for electrons. To maintain charge balance, these ions and electrons must therefore be transported from the locations where they are generated to the locations where they are consumed. For electrons this transport process is rather easy. As long as an electrically conductive path exists, the electrons will be able to flow from one electrode to the other. In the simple fuel cell in fig. 4, for example, a wire provides a path for electrons between the two electrodes. For ions, however, transport tends to be more difficult. Fundamentally, this is because ions are much larger and more massive than electrons. An electrolyte must be used to provide a pathway for the ions to flow. In many electrolytes, ions move via "hopping" mechanisms. Compared to electron transport, this process is far less efficient. Therefore, ionic transport can represent a significant resistance loss, reducing fuel cell performance. To combat this effect, the electrolytes in technological fuel cells are made as thin as possible to minimize the distance over which ionic conduction must occur.

Step 4: *Product Removal.* In addition to electricity, all fuel cell reactions will generate at least one product species. The H_2–O_2 fuel cell generates water. Hydrocarbon fuel cells will typically generate water and carbon dioxide (CO_2). If these products are not removed from the fuel cell, they will build up over time and eventually "strangle" the fuel cell, preventing new fuel and oxidant from being able to react. Fortunately, the act of delivering reactants into the fuel cell often assists the removal of product species out of the fuel cell. The same mass transport, diffusion, and fluid mechanics issues that are important in optimizing reactant delivery (step 1) can be applied to product removal. Often, product removal is not a significant problem and is frequently overlooked. However, for certain fuel cells (*e.g.*, PEMFC) "flooding" by product water can be a major issue.

7. – Fuel cell performance

The performance of a fuel cell device can be summarized with a graph of its current-voltage characteristics. This graph, called a current-voltage (i-V) curve, shows the voltage output of the fuel cell for a given current output. An example of a typical i-V curve for a PEMFC is shown in fig. 6. Note that the current has been normalized by the area of the fuel cell, giving a current density (in amperes per square centimeter). Because a larger fuel cell can produce more electricity than a smaller fuel cell, i-V curves are normalized by fuel cell area to make results comparable.

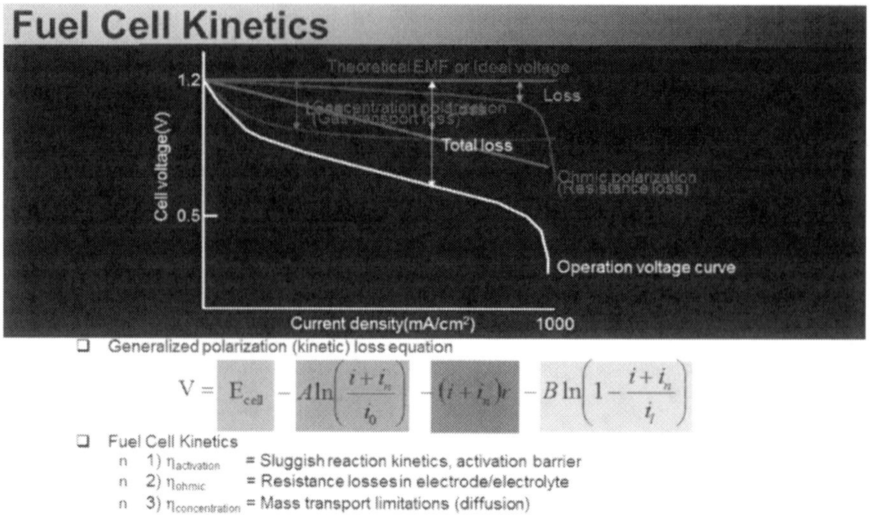

Fig. 6. – Schematic of fuel cell i-V curve. In contrast to the ideal, thermodynamically predicted voltage of a fuel cell (dashed line), the real voltage of a fuel cell is lower (solid line) due to unavoidable losses. Three major losses influence the shape of this i-V curve.

An ideal fuel cell would supply any amount of current (as long as it is supplied with sufficient fuel), while maintaining a constant voltage determined by thermodynamics. In practice, however, the actual voltage output of a real fuel cell is less than the ideal thermodynamically predicted voltage. Furthermore, the more current that is drawn from a real fuel cell, the lower the voltage output of the cell, limiting the total power that can be delivered. The power (P) delivered by a fuel cell is given by the product of current and voltage:

$$P = iV. \tag{6}$$

A fuel cell *power density curve*, which gives the power density delivered by a fuel cell as a function of the current density, can be constructed from the information in a fuel cell i-V curve. The power density curve is produced by multiplying the voltage at each point on the i-V curve by the corresponding current density. An example of combined fuel cell i-V and power density curves is provided in fig. 7. Fuel cell voltage is given on the left-hand y-axis, while power density is given on the right-hand y-axis.

The current supplied by a fuel cell is directly proportional to the amount of fuel consumed (each mole of fuel provides n moles of electrons). Therefore, as fuel cell voltage decreases, the electric power produced *per unit of fuel* also decreases. In this way, fuel cell voltage can be seen as a measure of fuel cell efficiency. In other words, you can think of the fuel cell voltage axis as an "efficiency axis." Maintaining high fuel cell voltage, even under high current loads, is therefore critical to the successful implementation of the technology.

 DOI: 10.1051/epjconf/201714800013

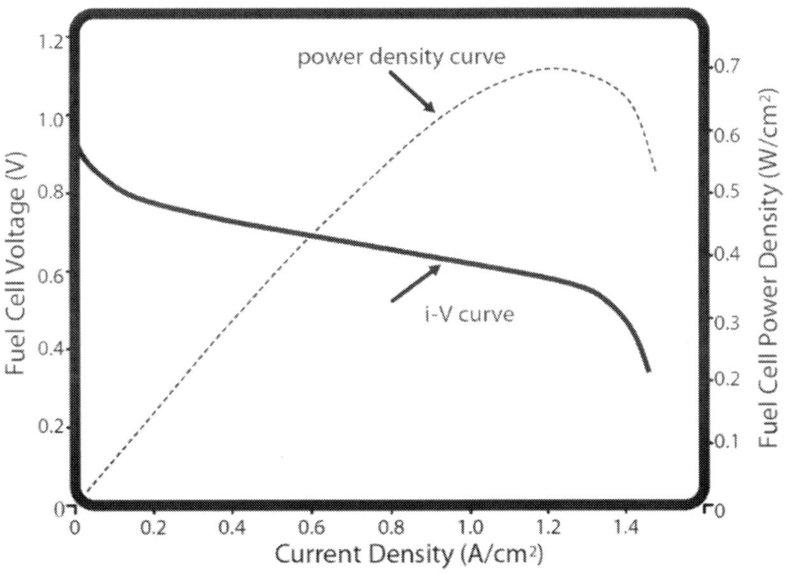

Fig. 7. – Combined fuel cell *i-V* and power density curves. The power density curve is constructed from the *i-V* curve by multiplying the voltage at each point on the *i-V* curve by the corresponding current density. Fuel cell power density increases with increasing current density, reaches a maximum, and then falls at still higher current densities. Fuel cells are designed to operate at or below the power density maximum. At current densities below the power density maximum, voltage efficiency improves but power density falls. At current densities above the power density maximum, both voltage efficiency and power density fall.

Unfortunately, it is hard to maintain a high fuel cell voltage under current load. The voltage output of a real fuel cell is less than thermodynamically predicted voltage output due to irreversible losses. The more current that is drawn from the cell, the greater these losses. There are three major types of fuel cell losses, which give a fuel cell *i-V* curve its characteristic shape. Each of these losses is associated with one of the basic fuel cell steps discussed in the previous section:

1. Activation losses (losses due to electrochemical reaction).
2. Ohmic losses (losses due to ionic and electronic conduction).
3. Concentration losses (losses due to mass transport).

The real voltage output for a fuel cell can thus be written by starting with thermodynamically predicted voltage output of the fuel cell and then subtracting the voltage drops due to the various losses:

(7) $$V = E_{\text{thermo}} - \eta_{\text{act}} - \eta_{\text{Ohmic}} - \eta_{\text{conc}},$$

where

V = real output voltage of fuel cell,

E_{thermo} = thermodynamically predicted fuel cell voltage output; set by the laws of thermodynamics,

η_{act} = activation losses due to reaction kinetics,

η_{Ohmic} = Ohmic losses from ionic and electronic conduction,

η_{conc} = concentration losses due to mass transport.

The three major losses each contribute to the characteristic shape of the fuel cell i-V curve. As shown in figs. 6 and 7, the activation losses mostly affect the initial part of the curve; the Ohmic losses are most apparent in the middle section of the curve, and the concentration losses are most significant in the tail of the i-V curve.

8. – Fuel cell technology

Now that we have briefly overviewed the basic working principles of fuel cells, we will conclude our short introduction of fuel cells with a brief discussion of fuel cell technology.

The first step in the fabrication of a commercial fuel cell device is the production of individual fuel cell "membrane electrode assembly" units or "MEAs". A fuel cell MEA, which consists of an anode electrode, an electrolyte membrane layer, and a cathode electrode, is the basic repeat unit of a fuel cell device. Depending on the type of fuel cell, MEA fabrication can differ significantly. The most common fuel cells are polymer fuel cells (PEMFCs) and solid oxide fuel cells (SOFCs). Figure 8 illustrates the typical steps involved in the fabrication of both PEMFC and SOFC MEAs.

As we have learned, the voltage of a single fuel cell is limited to about 1 V. Furthermore, under load, the output voltage of a single hydrogen fuel cell is typically 0.7–0.8 V. This range generally corresponds to an operational "sweet spot" where the electrical efficiency of the fuel cell is reasonable (around 65%) and the power density of the fuel cell is near its maximum. However, most real-world applications require electricity at several, tens, or even hundreds, of volts. How do we get 0.7 V fuel cells to supply the high-voltage requirements of real-world applications? One option is to interconnect multiple fuel cells in series. Connected in series, fuel cell voltages sum. This technique, known as fuel cell "stacking," permits fuel cell systems to meet any voltage requirement.

In addition to building voltage, fuel cell stacks are often designed with these goals in mind:

- Simple and inexpensive to fabricate.
- Low-loss electrical interconnects between cells.
- Efficient manifolding scheme (for reactant gas distribution).
- Efficient cooling scheme (especially for high-power stacks).
- Reliable sealing arrangements between cells.

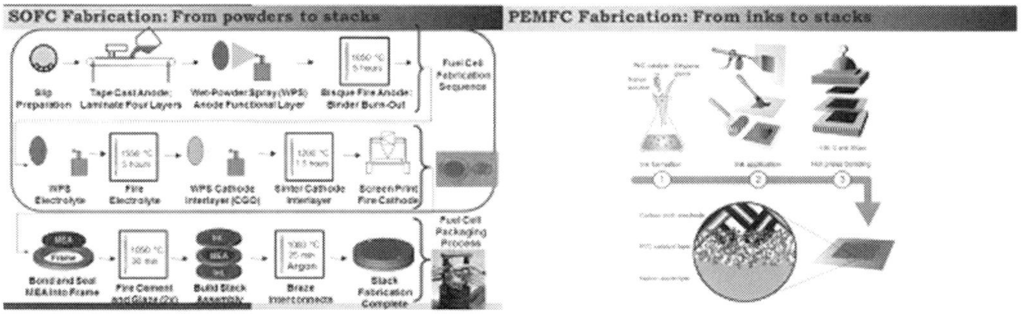

Fig. 8. – Schematic illustration of the processes involved in fabrication SOFC and PEMFC MEAs.

Figure 9 illustrates a common form of fuel cell stacking, referred to as *vertical* or *bipolar plate* stacking. In this configuration, a single conductive flow structure or plate is in contact with both the fuel electrode of one cell and the oxidant electrode of the next, connecting the two fuel cells in series. The plate serves as the anode in one cell and the cathode in the next cell, hence the name *bipolar plate*. Bipolar stacking is similar to how batteries are stacked on top of one another in a flashlight. Bipolar stacks have the advantage of straightforward electrical connection between cells and exhibit extremely low Ohmic loss due to the relatively large electrical contact area between cells. The bipolar plate design leads to fuel cell stacks that are robust. Most conventional fuel cells stacks adapt this configuration, although other stacking alternatives such as tubular or segmented-in-series designs are also possible.

Beyond simply connecting multiple fuel cell MEAs together in a stack, many other components are required to make a complete fuel cell system. These components include

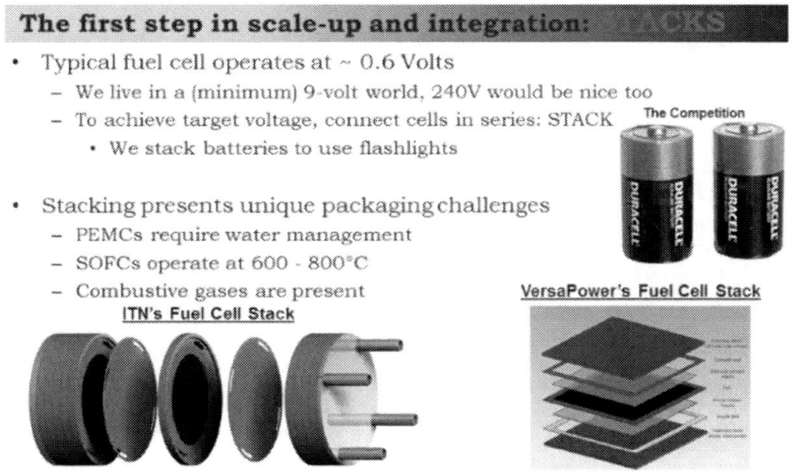

Fig. 9. – Schematic illustration of fuel cell stack design.

DOI: 10.1051/epjconf/201714800013

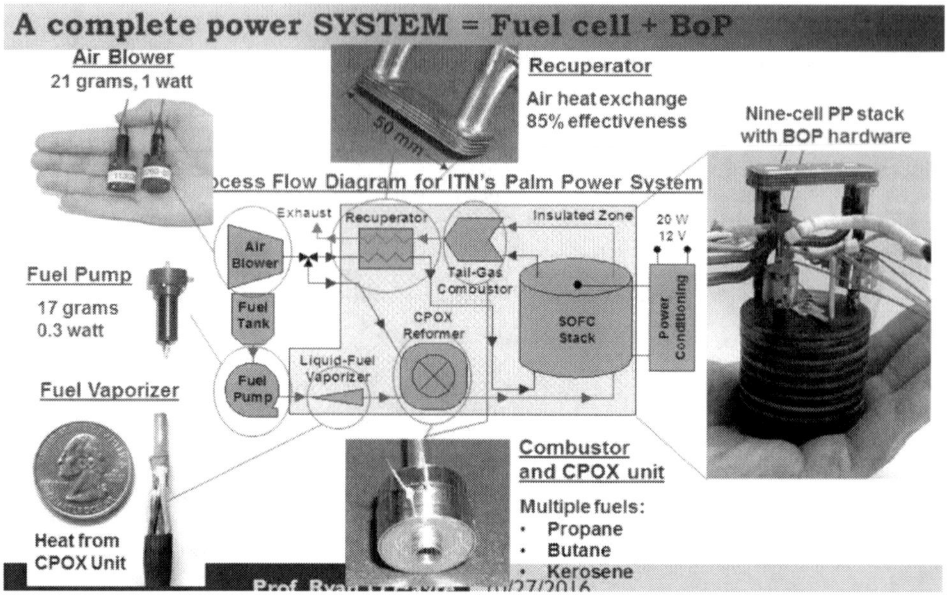

Fig. 10. – Schematic example of the various balance-of-plant components making up a complete portable solid oxide fuel cell system.

devices that provide the fuel supply, cooling, power regulation, and system monitoring, to name a few. Often, these devices, which are collectively known as the *Balance of System*, or *Balance of Plant* can take up more room (and cost) than the fuel cell unit itself. Those that draw electrical power from the fuel cell are called ancillaries, or parasitic power devices.

The target application strongly dictates fuel cell system design. In utility-scale stationary power generation, where reliability and energy efficiency are at a premium, there is a strong incentive to include beneficial system components. In portable fuel cell systems, where mobility and energy density are at a premium, there is a strong incentive to minimize system components.

These most common subsystems present in most fuel cell devices include:

- The fuel cell subsystem.
- The thermal management subsystem.
- The fuel delivery/processing subsystem.
- The power electronics subsystem.

Figure 10 provides an example of the type of balance of plant components that might make up a complete portable solid-oxide fuel cell system. Examples of both a large-scale (220 kW) and a small (50 W) portable fuel cell system are shown in fig. 11. In both examples, the fuel cell stack itself takes up less than 30% of the total size of the system.

DOI: 10.1051/epjconf/201714800013

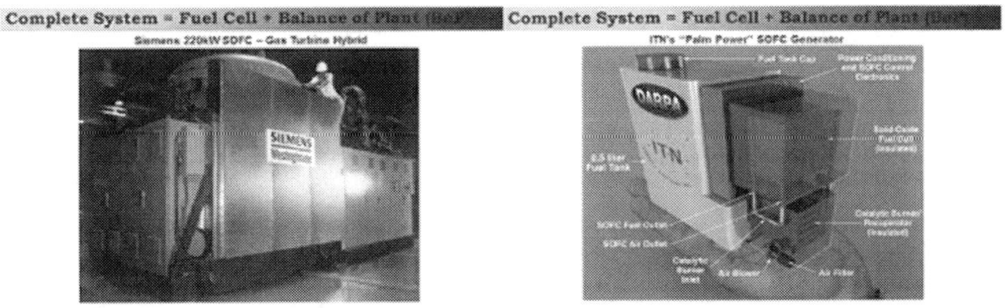

Fig. 11. – Examples of large-scale (220 kW) and small-scale (50 W) fuel cell systems.

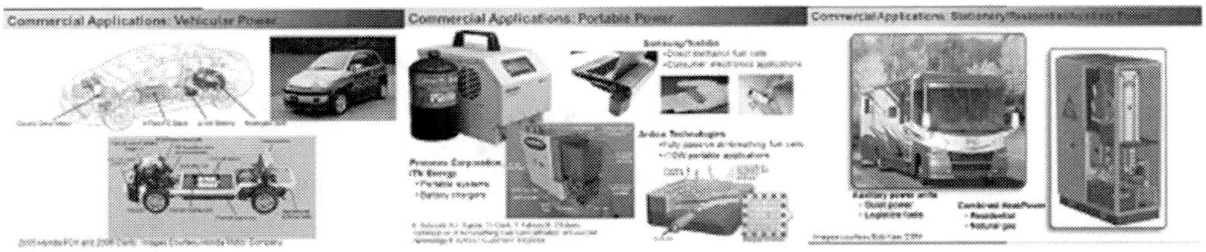

Fig. 12. – Primary fuel cell market opportunities include vehicular, portable, and stationary/distributed power applications.

As illustrated in fig. 12, fuel cells are primarily being considered for use in three main application areas:

- Vehicular applications.
- Stationary/distributed generation applications.
- Portable applications.

Each application area has its own unique requirements and challenges. Although fuel cells have received the most public attention for vehicular applications, in some ways this represents the most challenging market application for fuel cells due to the need for low cost, high durability, extremely high power density, and rapid start-stop cycle capability. Most fuel cell experts believe that portable and stationary (*e.g.* distributed generation or back-up power) applications offer more attractive near-term opportunities for fuel cell commercialization.

Figure 13 provides two examples of current markets where fuel cells have garnered significant recent success: 1) backup power for remote electronic installations (such as cell-phone towers), and 2) fuel cell powered forklifts for industrial warehouse facilities. In each case, fuel cells offer unique and compelling advantages not available from competing power technologies such as batteries or diesel engines.

Fuel cell technology is truly still in its infancy, and there are significant research challenges that must still be overcome before fuel cells are widely implemented. Decreasing

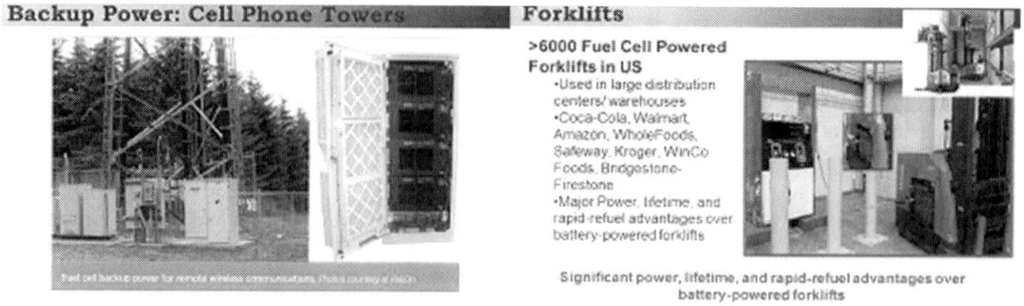

Fig. 13. – Backup power for remote electronic installations (such as cell-phone towers) and fuel-cell powered forklifts for warehouse facilities represent two niche markets where fuel cells have gained significant traction.

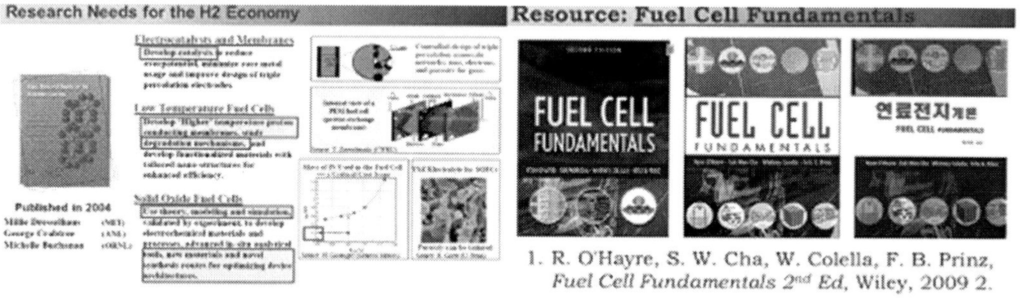

Fig. 14. – Suggested resources for further information of the fundamentals of fuel cell science and technology as well as the current major research challenges and opportunities facing the field.

cost, increasing lifetime/durability, and expanding the ability of fuel cells to operate on multiple fuels (beyond hydrogen) represent some of the most important broad-based challenges in the fuel cell research community. For further information on fuel cell science and technology and the research challenges and opportunities facing this field, readers are invited to consult the additional references provided in fig. 14.

EPJ Web of Conferences **148**, 00014 (2017) DOI: 10.1051/epjconf/201714800014
LNES 2016

Thermal energy storage for CSP (Concentrating Solar Power)

Xavier Py(*), Najim Sadiki, Régis Olives, Vincent Goetz and Quentin Falcoz

Laboratoire PROcédés Matériaux et Energie Solaire PROMES-CNRS UPR 8521
University of Perpignan - Via Domitia, Perpignan, France

Summary. — The major advantage of concentrating solar power before photovoltaic is the possibility to store thermal energy at large scale allowing dispatchability. Then, only CSP solar power plants including thermal storage can be operated 24 h/day using exclusively the solar resource. Nevertheless, due to a too low availability in mined nitrate salts, the actual mature technology of the two tanks molten salts cannot be applied to achieve the expected international share in the power production for 2050. Then alternative storage materials are under studies such as natural rocks and recycled ceramics made from industrial wastes. The present paper is a review of those alternative approaches.

1. – General context

The so-called energy transition is based on a worldwide extended growing of renewable energies. This concerns all kinds of renewable energies leading to local mix adapted to the local resources and needs. According to the fact that almost all renewable energies

(*) Email: py@univ-perp.fr (corresponding author)

present major intermittencies (especially solar and wind energies) and mismatch with the consumers needs, numerous energy storage units will be also highly needed.

Then, among the major issues associated to the energy transition, the question of the availability in mineral resources to build all the corresponding energy processes and energy storage facilities could lead to major bottlenecks. More generally, the LCA (life cycle analysis) should be addressed properly for each involved process to assess its environmental and energetic benefits and corresponding environmental paybacks. This concerns the embodied energy, the green house gas, the water and ecotoxicity impacts, the risks in conflict of use, etc.

Surprisingly, even for the renewable energies technologies developed for almost half a century, their associated LCA are very recent and still scarce. Very often, those technologies are viewed as just usual industrial facilities while their contribution to a sustainable development has to be addressed not only in terms of economy but also in terms of environment and social benefits.

The present paper is focused on the particular case of thermal energy storage (TES) materials for concentrating solar power (CSP) as an illustration of the above more general concern.

2. – CSP technologies

The production of power from the solar radiation is currently achieved by two very different routes: 1) the very popular photovoltaic technique allowing the conversion of the whole solar radiation directly into electricity and 2) the less known thermal route based on a first conversion of the solar radiation (only the direct fraction) to heat to produce steam which is turbined such as in all other thermal power plant [1].

The major advantage of the second one is the fact that it is much easier and more affordable to store the energy at large scale in the form of heat than in the form of electricity. Therefore, the photovoltaic power stations produce rather erratic electricity to the grid inducing major risks of stability and the corresponding economic value of the product is necessarily low. In comparison, thanks to large scale thermal storage, solar thermal power plants can produce 24 h/day very stable power levels to the grid such as all kinds of other thermal power plants. This is the case of Gemasolar (Spain 2011) or Crescent dune (USA 2015) solar power plants implemented recently which are able, thanks to 15 h of TES capacity, to produce 20 and 110 MW of constant power to the grid, respectively.

Other CSP plants such as ANDASOL have less extended TES units, here 7.5 h of capacity, allowing already the possibility to balance properly the solar radiation variation daily. Their production can be easily extended to the night by using fossil backup. Other CSP plants such as the historic CEGS (USA), ran for decades without storage and using only fossil backup.

All those CSP plants are based on two families of concentrating technologies gathering 1) the linear concentrators such as Linear Fresnel or solar trough mirrors (fig. 1) and 2) central receiver plants (fig. 2) such as tower CSP or dish Stirling (fig. 3) facilities

Fig. 1. – Linear concentrating techno of solar trough.

Fig. 2. – Central receiver concentrating techno.

Fig. 3. – Dish Stirling.

Fig. 4. – Two-tank molten salt TES for CSP.

(those last ones are more devoted to remote areas, small or medium range of power and do not include TES). The former ones operate up to 400 °C while the second ones operate currently up to 800 °C and could even work at higher temperature (they are still limited to the availability of high-temperature materials).

3. – Current thermal storage for CSP

Up to date, all the TES sub-units of industrial CSP plants devoted to several hours of production are based on a unique technology initially developed in Themis-France. This technology [2], the so-called "two-tank molten salt" one (fig. 4), is composed of two separated tanks in which all the molten salt (in all cases a mixture of nitrate salts) used simultaneously as heat transfer fluid (HTF) and thermal storage material can be stored at the two temperature levels. The molten salt stores the thermal energy by change of its temperature, the so-called sensible heat approach. During the sunniest hours of the day, the intentionally over-scaled solar field of the plant delivers a heat power shared between the power block and the TES unit. During the period of time during which the output heat power from the solar field is lower than the nominal feeding power of the power block, the TES unit contribute to balance it. Therefore, using the TES unit, the power block of the plant is fed at nominal power and delivers a constant electric power to the grid. This approach allows also to reduce the maintenance cost of the power block and to increase its life time expectancy.

In this process, the salt has to be kept under liquid state only. As the basic binary eutectic "solar salt" made of $NaNO_3$ and KNO_3 presents a solidification temperature of 250 °C, the different equipments have to be maintained above this critical temperature to avoid any plug. Soon during the eighties, major efforts were done to find other molten salt mixtures of lower solidification temperature. It was the case in Themis in which a mixture of $NaNO_2/KNO_3/NaNO_3$ with a solidification temperature of 142 °C was used. This issue is still under study and other mixtures are proposed in the literature.

The major current efforts made on this technology are focused on the replacement of the two tanks by a unique one in which the hottest molten salt fraction is placed

DOI: 10.1051/epjconf/201714800014

over the coldest fraction. In this "thermocline" approach [3], the two temperature levels are naturally separated by the difference in density. Its major advantage is to reduce the overall cost of the TES unit (by about 30%) by using a single tank. Nevertheless, heat transfer by diffusion at the interface between the two temperature levels reduces the thermal efficiency. During the charging period, the coldest salt fraction is pumped at the bottom of the tank to be heated by the solar field through a heat exchanger and to go back to the top of the TES tank. During the discharge step, the hottest salt fraction is pumped at the top of the tank to feed the power block and the resulting coldest salt is fed back at the bottom of the tank. Therefore, during those two operating steps, the thermal gradient zone of the salt in the tank moves up (under charge) and down (during the discharge). Efforts are made to develop floating insulating membranes of intermediate density following the thermocline during the charge/discharge periods.

Few experiences have been made to use thermal oil as HTF and TES material (SEGS-1 USA) associated to a mixture of granite gravel and sand as filler (to reduce the needed amount of oil in the tank). Unfortunately, this storage unit was destroyed by a major fire inducing a critical environmental impact and a significant societal traumatism. This attempt was the first and last industrial oil-based TES in a CSP plant.

The above TES technology for CSP developed during the eighties for the pilot plant central receiver approach (Themis France and Solar-Two USA) has been used as standard TES system in all industrial CSP plants built from the beginning of the 21st century. This is particularly the case of solar trough CSP plants acknowledged at this time as the most mature CSP approach (but without integrated TES sub-system) to which the mature TES unit of central receiver CSP plants has been added. The hybrid result is illustrated by the ANDASOL (2009, 50 MWe, 7.5 h of TES, Granada Spain) plant which can be considered today as a standard of the technology. More recently, central receiver industrial plants of Gemasolar (2011, 20 MWe, Spain) and Crescent-Dune (2015, 110 MWe, USA) offer 24 h/day of solar only operation at constant power due to their 15 h of TES capacity.

4. – Major limitations and opportunities

The two-tank molten salt approach described in sect. **3** can be considered as mature and presents actually almost all the expected performances. Nevertheless, this TES technology applied to the worldwide CSP growing is not sustainable.

A first aspect which has been considered only very recently (surprisingly for a technology developed during the eighties) is the environmental impact of the approach. The first complete LCA (Life Cycle Analysis) of CSP solar trough plant including the TES unit has been published in 2011 [4]. This very detailed study highlights the fact that the current TES technology represents 19.3% of the GHG (green house gas) and 18% of the CED (cumulative energy demand) contents of the ANDASOL-like plant.

In terms of costs, the TES unit represents about 9% of the total initial investment and the molten salt 49% of the TES cost.

Therefore, the economical and environmental contributions of the TES unit (which is of 7.5 h of capacity in those studies while 24 h/day running CSP plants involve 15 h of TES capacity) are very significant and their reduction can be already considered as strategic.

Another major concern is the availability of the involved nitrate salts. According to the IEA analysis, the expected 10% of CSP power contribution in the worldwide electricity production in 2050 could induce major tension and even conflict of use in the mineral materials markets. The deep analysis of the mineral resources involved in a CSP plant equipments highlights the fact that only the nitrate molten salt should present major concern in term of availability [5, 6].

Currently, the nitrate salts $NaNO_3$ and KNO_3 of the TES for CSP are mainly (at about 60%) natural products mined in North Chile, the other part is from the chemical industry. Even if the major part of this solar salt is a natural mineral material, it presents a significant environmental footprint due to the mine extraction, purification, shipping, implementation in TES. In terms of availability, Chile provides historically salts from the mid of the 19th century with a peak of production at the first world war at a rate of about 3 Mt/y. Due to the needs in nitrates for the war industry, Germans built at that time the industrial synthetic way to produce nitrate salts, inducing a major destabilization in the worldwide market. Today, Chile produces about 0.8 Mt/y of nitrate salts for a market shared between the agriculture (as fertilizer) and the chemical industry. According to the fact that the CSP grow needed to achieve the 10% expected share in the worldwide electricity production of 2015 corresponds to a need in nitrate salts of about 20 Mt/y, this current availability is critically too low. Moreover, the use of the nitrate salts to produce electricity could not be considered as a priority before the agriculture.

Therefore, there is a critical need in alternative HTF and TES materials to achieve the expected CSP grow of the next decades.

5. – Alternative approaches

About one decade ago, the worldwide CSP academic and industrial community has identified the lack of availability and the environmental footprint of the nitrate salts. From this time, major efforts have been made to find alternative technologies based on two major approaches: one trying to reduce the amount of salts of the current processes, the other one trying to develop new technologies without salts.

The former one is for instance focused on the development of filler storage materials to be placed in the TES tank to replace as much as possible the volume of salt [3]. In this case, the molten salt is mainly reduced to its HTF functionality and its needed amount decreased by about 75%. Another emerging and very promising way is based on nanofluids [7]. By adding just few %wt of nano sized solid particles (such as TiO_2 or SiC or nano carbonaceous materials), the sensible heats storage capacity and the thermal conductivity of the molten salt can be increased by 30 to 50%. If this effect on the thermal conductivity was already known for decades, the effect on the thermal capacity has been discovered just few years ago. The corresponding fundamental mechanisms responsible for this C_p increase are still not fully understood and the transfer of this approach to the industrial scale will have to be addressed also. Nevertheless, this new subject opens a new route to reduce the needs in molten salt juts by increasing their effective performances.

The other way to overcome the lack of nitrates availability is to switch to other HTF and TES materials. This is for instance the case of the proposed CSP central receiver tower plants in which hot air is used as HTF and sand or rock or high-temperature ceramics as TES materials. Those promising approaches are still under pilot development.

5·1. *Natural materials.* – Natural materials to be used as TES materials have been considered from a long time. Already during the Paleolithic, men had to optimize their main heating technology of fire place. In some cases they had only fast fuels available (like straw) and simultaneously needs of long cooking durations. Academics have found evidence of use of rocks as TES materials to store heat of straw or other fast-burning fuel for subsequent cooking on the heated rocks. Others were confronted to the incompatibility between fire and water to produce steam (which is obvious for us but was certainly less obvious at the beginning of the fire domestication). They developed an iterative and then very smart technique: rocks were heated in a fire place to be soaked in a water pool next to the fire place and when the rock was cooled down it was moved back to the fire place, etc., and the process reproduced up to the boiling of the water volume.

Still during the Paleolithic, flint stones were used to produce cutting tools. If the corresponding technique of flint stone shaping is well known, the fact that some flint stones were intentionally heated for an easier shaping is less known today. Actually, some flint stones are difficult and even impossible to shape while after controlled heat treatment the transformation is possible and the final product very valuable. As a direct flint heating on a fire place induces an irremediable breaking of the piece, men had to develop very sophisticated techniques to allow a uniform heat treatment in the stone volume. Today, this knowledge has been lots but the observed remaining tools demonstrate the possibility to find heat treatments allowing the stabilization of stones to be used at high temperature without breaking.

For years, instead of using costly synthetic high-temperature ceramics, the CSP community have tried to used available rocks as TES materials. Today, as the environmental and economical issues are more and more considered during the CSP process design and development, this alternative approach is still under focus.

We can consider two main kind of rocks for this thermal applications: those presenting almost no structural or chemical change during the first heating (such as granite or basalt) and those presenting major modifications during the first heating. The former ones can be potentially used directly while the second ones have to be first stabilized by a proper heat treatment before being used at high temperature.

As an illustrative example, basalt samples from France and Egypt (fig. 5) and flint stone from France (fig. 6) have been characterized and tested from room temperature up to 800–1000 °C (unfortunately, stone properties are usually given in the literature at room temperature but very scarcely at high temperatures).

 DOI: 10.1051/epjconf/201714800014

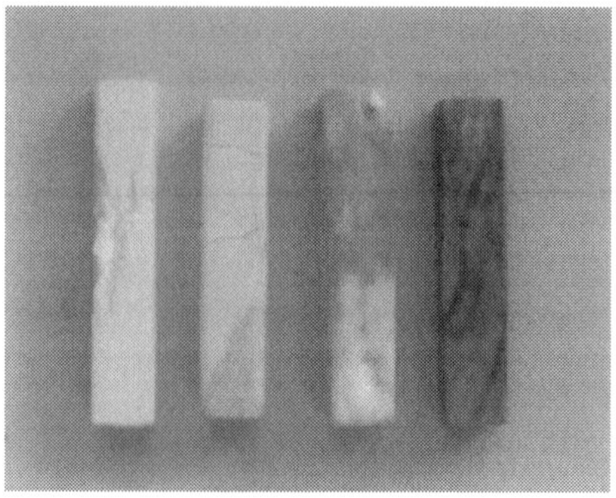

Fig. 5. – Basalt samples from France and Egypt.

Fig. 6. – Flint stone sample.

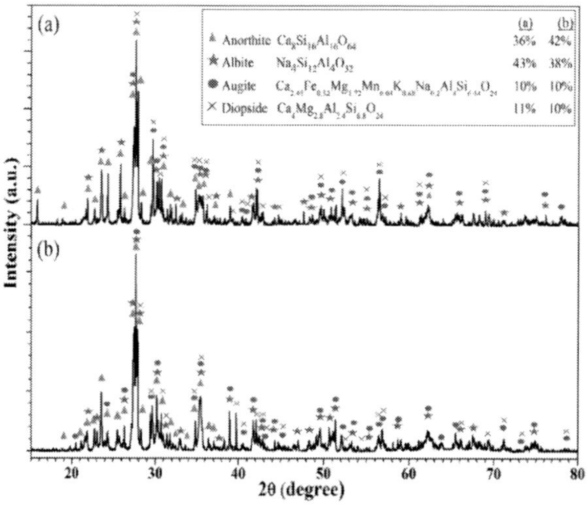

Fig. 7. – XRD diagrams of basalt stones before (a) and after (b) heat treatment at 1000 °C.

DOI: 10.1051/epjconf/201714800014

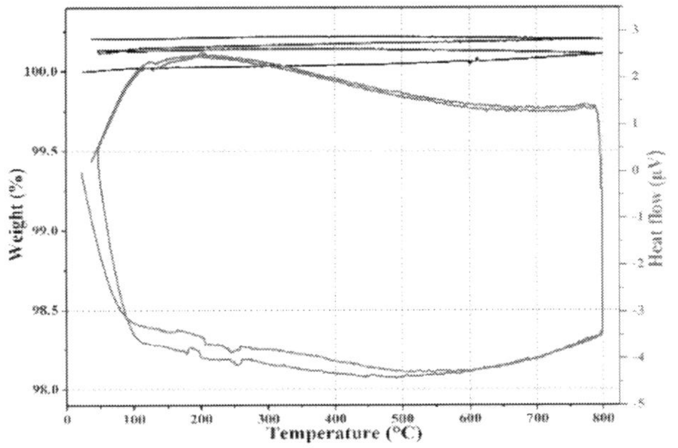

Fig. 8. – DSG and TGA curves of basalt up to 800 °C.

As shown in fig. 7, the basalt stones present very stable XRD diagrams demonstrating the fact that those stones offer very stable crystal structure under heating up to 1000 °C. Corresponding DSG and TGA characterizations have been made up to 800 °C (the highest CSP temperature today). As illustrated in fig. 8, the DSG thermal cycles demonstrate the fact that the stone presents sensible heat effect only without any king of other thermal signal which could be attributed to chemical reaction or other major transformation. The TGA curves under thermal cycling highlights a very small change in sample weight due to the departure of a slight fraction of water from the stone.

In terms of thermophysical properties considered for high-temperature TES applications, the storage capacity of basalt stones range between 750 to 1100 J/(kg K) from room temperature to 1000 °C (fig. 9). Those values are just those always observed for all kind of silica-based ceramics, natural (stones) or synthetic from the industry. This fact demonstrated that even if those materials are natural, their major properties present values in the expected range without major risk of variation. The thermal conductivity, is a very important property controlling the charge/discharge heating rate and consequently the power levels of the TES systems. It is currently balanced by the shaping of the storage material allowing to control the thickness of the storage modules and consequently the charge/discharge of heat deep to the very center of the material. A low thermal conductivity would have to be balanced by a corresponding extended specific external heat transfer interface inducing over costs of shaping and pressure drop on the HTF side. Consequently, a high thermal conductivity is advantageous. The thermal conductivities of the basalt stones from France and Egypt are illustrated in fig. 10 against temperature up to 700 °C. This demonstrates that variations in the range of 30% can be observed between two different samples depending upon their origin. As a matter of fact, those stones are not pure ceramics but have to be seen as composite ceramics made of a main matrix in which other crystals are distributed. Then, the overall observed thermal conductivity results from the contributions of the matrix, the different crystals (depending of their nature, sizes and shapes) and the crystals-matrix interfaces. It is obvious to consider

DOI: 10.1051/epjconf/201714800014

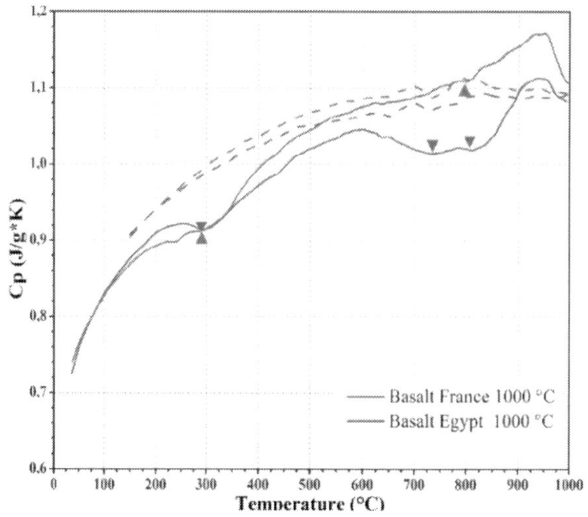

Fig. 9. – Thermal capacity of basalt stones under heating (continuous lines) and cooling (dashed lines).

that two basalts of different origins present significant differences in distributed crystals (in amount, nature sizes and shapes). Nevertheless, the range of value in thermal conductivity, between 1.5 to 2 W/(m K) falls within the usual range of silica-based ceramics without surprise. One other aspect could be considered as surprising: it is the fact that those thermal conductivities increase with temperature. Usually, the thermal conductivity of solid materials falls with temperature. Here, the observed behavior results from not only the effect of temperature on the intrinsic thermal conductivity of the involved

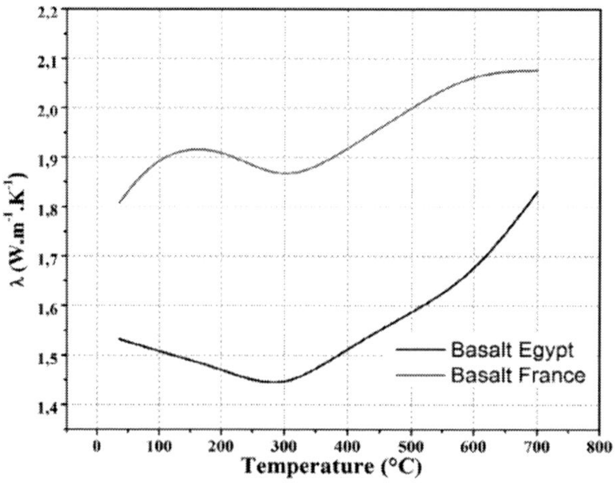

Fig. 10. – Thermal conductivity of basalt stones.

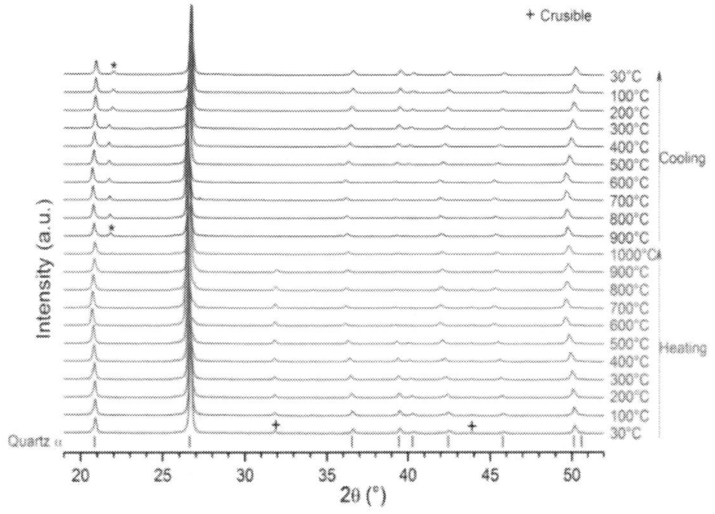

Fig. 11. – XRD diagrams in temperature of flint stones.

ceramics but also on the thermal resistances which are the interfaces between the dispersed crystals and the main matrix. Under heating, when the thermal expansions of the crystals are higher than that of the matrix, the interfacial gaps are progressively closed and consequently the thermal resistance decrease. This results in an observed increase in thermal conductivity.

Those results associated to the fact that the observed heated samples were not broken after treatment, validate the use of basalt stones for high-temperature applications such as TES for CSP.

In the other side of stone behavior, four samples of flint (fig. 6) representative of the four main flint stone families have been experimented and characterized.

In terms of stability (see fig. 11), the XRD diagrams seems to demonstrate no change. In fact, the crystal structure does not change in nature but during the first heating the flint experiences major modifications inducing corresponding thermal signal and weight variations observed in fig. 12. At low temperature (about 100 °C), the free water departure induces a weight loss and a endothermic peak. From 400 to 600 °C, silanol species SiOH are transformed in Si-O bounds which participate to the stabilization of the material:

$$\text{SiOH} + \text{SiOH} \rightarrow \text{Si-O-Si} + H_2O. \quad (1)$$

Nevertheless, the corresponding local formation of H_2O can induce major local formation of pressurized steam due to a lack in open porosity. Consequently, a too fast heating rate can induce the formation of unrealized high-pressure steam volumes and to the breaking of the stone. A much more controlled slow heating rate could offer the possibility to extract the formed steam through the porosity. Anyway, those results highlights the fact

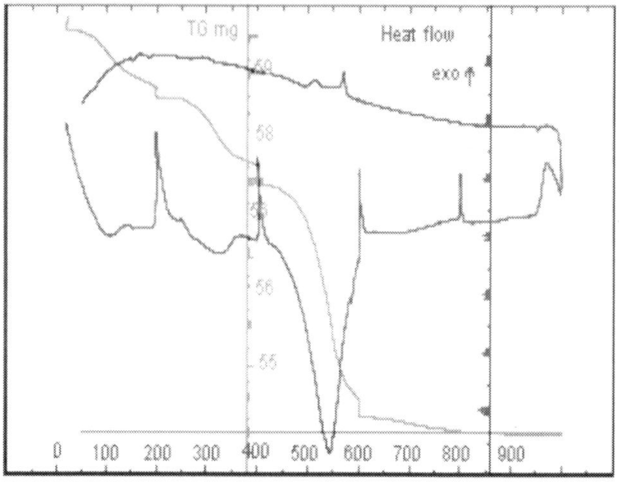

Fig. 12. – DSG diagram of flint stone up to 1000 °C.

that those stone are better to be used in medium range concentrating CSP technologies (linear Fresnel and solar trough) below 400 °C while basalt stone can be used in all CSP technologies even the future high temperatures of 800–1000 °C.

In the thermal range of linear concentrating CSP technologies (below 400 °C), the thermophysical properties of the four flint samples have been characterized.

As illustrated in fig. 13, the thermal capacity under heating and cooling ranges in the usual range of silica materials with a rather conventional behavior. In terms of thermal

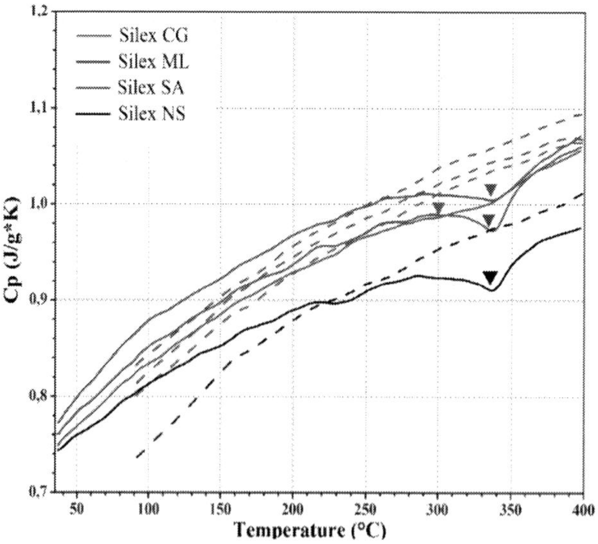

Fig. 13. – Thermal capacity of flint stones up to 400 °C.

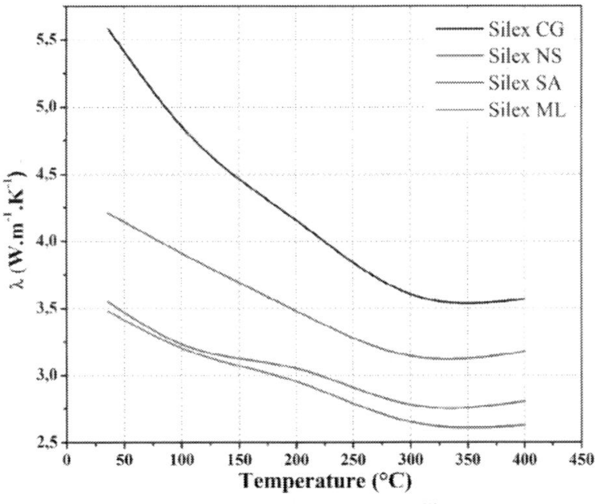

Fig. 14. – Thermal conductivity of flint stones up to 400 °C.

conductivity (fig. 14), the observed values (between 2.5 and 5.5 W/(m K)) are larger than those previously observed for the basalt. Those results show also a significant difference between the four samples, highlighting the effect of the composite nature of those natural product. Moreover, comparatively to the opposite behavior of the basalt, the thermal conductivities of all the experienced flint stones decrease with temperature.

Those two illustrative examples of very stable stones (basalt) and transforming stones (flint) show simultaneously the high potential of those materials for high-temperature TES applications but also their limitations and the critical need in systematic preliminary characterization before use.

In another aspect but very important for the TES application, the compatibility between those solid fillers and the HTF currently used has to be addressed properly. This has been done with hot air up to 1000 °C, with thermal oils up to 400 °C and nitrate molten salts. One major difficulty concerning this issue is the fact that those compatibility tests should be done during very long periods of times.

5·2. *Recycled materials.* – Considering the above limitations and bottlenecks related to TES material availability, cost and footprints, a new approach has been proposed a decade ago based on recycling inorganic industrial wastes.

This concerns asbestos-containing wastes (ACW) [8], fly ash (FA) [9, 10] and bottom ash (BA), slags from metallurgy [11] and all other mineral-like industrial wastes [12].

The first major advantage of this mineral resource is its availability. For instance, coal-fired power plants are responsible of a worldwide production of about 750 Mt/y of fly ash (micron-size spherical silica particles) released in their fumes and gathered by electrostatic filters. This first example compared to the annual CSP needs of some 20 Mt/y demonstrates the high availability of those inorganic wastes.

Another advantage is the fact that some of those wastes are very dangerous. For

Fig. 15. – ACW glass obtained by fast cooling.

instance, asbestos-containing wastes are responsible for severe diseases and the only way to inert them irreversibly is to melt them to destroy their micro-fibrous structure. The problem is that this melting treatment has to be done at 1400 °C and consequently presents a high GHG and CED content. The only way to offer a positive environmental balance of the obtained inert product is to use it as high-temperature storage material. Another great advantage here is the fact that the melting step offers the possibility to control the cooling period and consequently the structure and properties of the obtained ceramic. Moreover, this cooling step can be made in a mould to produce a shaped TES module allowing an efficient heat transfer with the flowing HTF used to charge/discharge the material.

The particular case of metallurgic slags is even better: as slags are directly produced under molten state at the outlet of the metallurgic furnaces, the TES materials and modules can be directly made without additional heat treatment.

Concerning Fly Ashes, as they are in the form of micron-size spherical particles, they can be shaped in the form of TES modules by sintering, which is less costly (in terms of energy and financial costs) than melting.

Fig. 16. – ACW ceramic obtained by slow cooling or reheat of the corresponding glass.

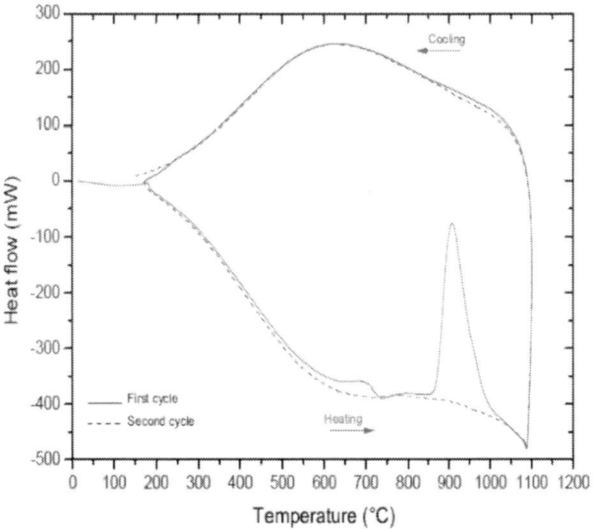

Fig. 17. – MSWI fly ash glass and ceramic DSG.

Among the possible reserve about this approach is the fact that, using industrial wastes as primary mineral resource, one could expect to get very variable properties of the obtained recycled ceramics. In fact, even if those materials contain a large amount of impurities, the four major components are always the same and they mainly impose the whole behavior and properties of the recycled ceramics.

As an example, municipal solid-waste incinerator ACW glass and ceramics have been studied. When the molten inert waste is cooled very fast, the matter has no time to form its crystal structure and we get a glass (fig. 15, an amorphous matter). If cooled at a lower rate, the material crystallizes and we get a ceramic such as in fig. 16.

Those two forms of recycled material have been characterized in terms of structure and properties. As illustrated in fig. 17 in the particular case of (MSWI) fly ash (similar results were obtained for the other wastes) under heat treatment, the DSG of the glass presents a vitrous transition peak at about 680 °C followed by a strong crystallization peak at 900 °C. The formed ceramic when cooled down to the room temperature and then submitted to a new full thermal cycle (dash lines) presents only sensible heat without any other effect. In terms of crystal structure, as illustrated in the XRD diagrams of fig. 18 the glassy material presents an amorphous flat curve, first crystals appear at about 900 °C confirming the DSG results and the other diagrams made on the ceramic under thermal cycling demonstrate a very stable crystal structure.

Those results highlight a very stable thermal behavior of those recycled ceramics up to more than 1000 °C.

In terms of thermophysical properties, the thermal capacity of all the obtained ceramics falls in the usual range (from 800 to 1100 J/(kg K)) of silica ceramics and rocks. This is shown in figs. 19 and 20 for ACW and MSWI ceramics from room temperature up to 1000 °C.

DOI: 10.1051/epjconf/201714800014

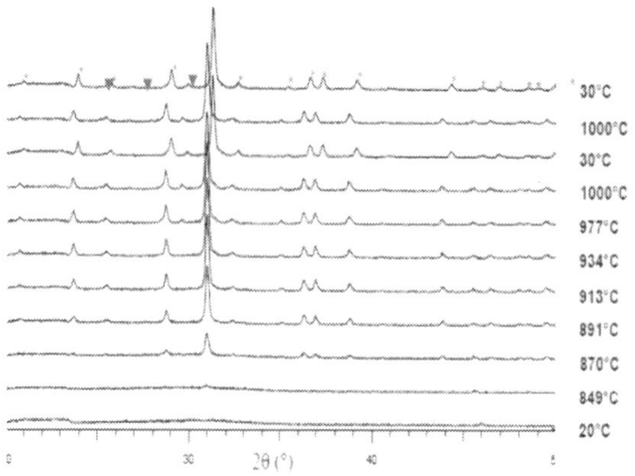

Fig. 18. – MSWI FA glass and ceramic XRD under T.

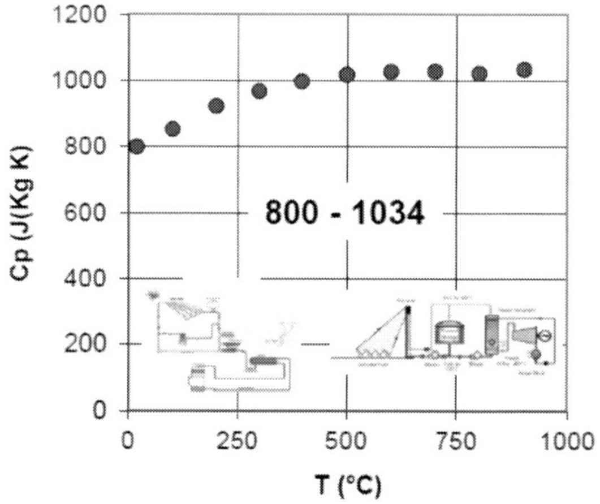

Fig. 19. – Thermal capacity of ACW ceramic.

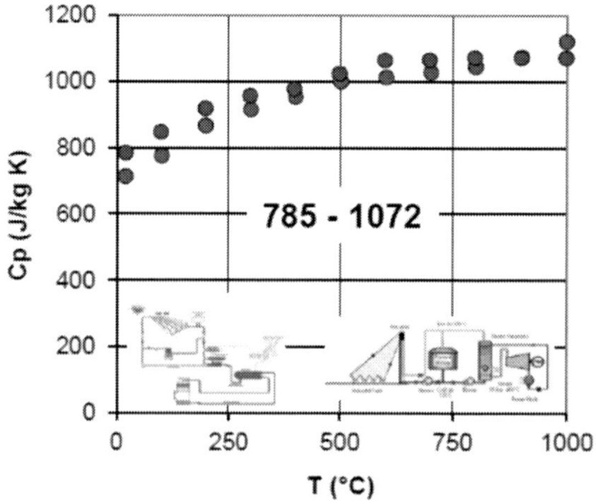

Fig. 20. – Thermal capacity of MSWI FA ceramic.

 DOI: 10.1051/epjconf/201714800014

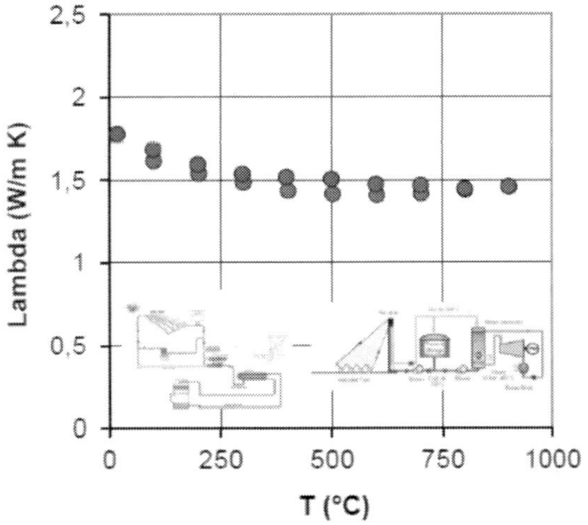

Fig. 21. – Thermal conductivity of ACW ceramic.

In terms of thermal conductivity, the two recycled ceramics present different behaviors. As illustrated in figs. 21 and 22, ACW ceramics present thermal conductivities in the range of 1.5 W/(m K) decreasing with temperature while MSWI ceramics present an increasing behavior for a similar range of values.

Therefore, the main thermophysical properties of the recycled ceramic are in the range of the expected values for high-temperature TES applications.

Another issue specifically critical for large scale industrial TES is the ability of the

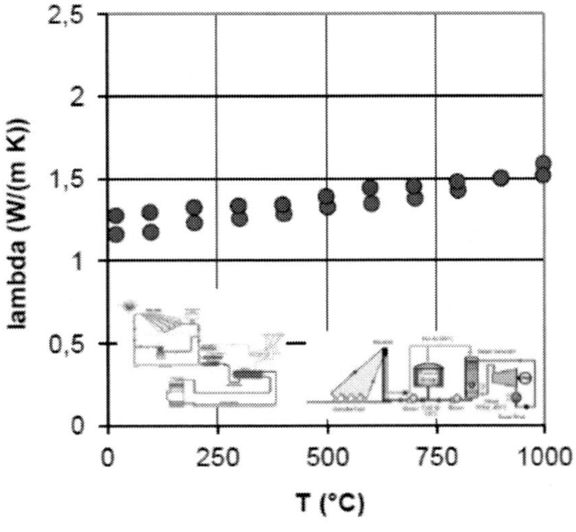

Fig. 22. – Thermal conductivity of MSWI FA ceramic.

Fig. 23. – Concentrating solar facility to test the ability of the ceramic to sustain thermal shocks.

storage materials to sustain repeated thermal cycling during 20 to 30 years between two levels of temperature [13], sometimes under pressure and in contact with a HTF for which the storage material has to offer a proper chemical compatibility. Obviously, those issue are very difficult to address at lab scale and during studies of reduced periods of time.

Nevertheless, corresponding works are achieved in labs and related industries.

As illustrative examples, the resistance to thermal shocks are tested in Promes-Odeillo under direct concentrated solar radiation (fig. 23) allowing severe heat treatments as in fig. 24 for which ACW ceramics were submitted to thermal cycles between 500 and 1000 °C under air with heating and cooling rates from 100 °C/min to 2500 °C/min [13].

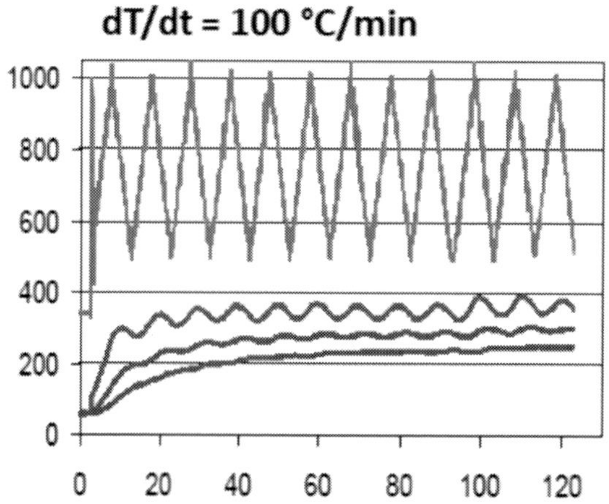

Fig. 24. – Experimental temperature at the surface of the ceramic and at different depth.

DOI: 10.1051/epjconf/201714800014

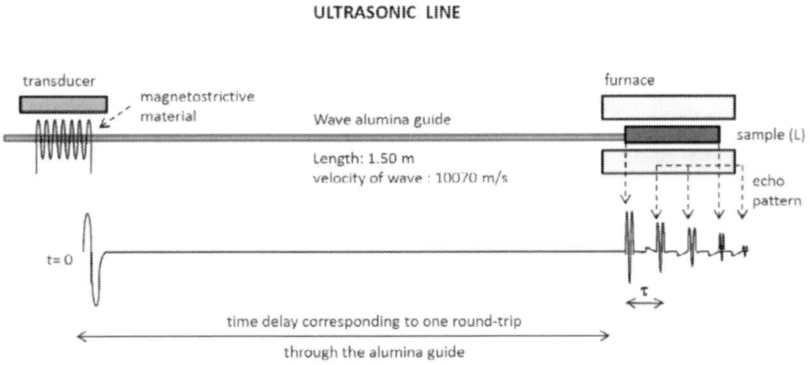

Fig. 25. – Ultrasonic equipment for the measurement of Young modulus under temperature.

The ceramic material was able to sustain those extreme treatments without failure and consequently they will be able to sustain the much smarter usual conditions of industrial TES.

In terms of thermomechanical properties, the Young modulus and dilatometry of the recycled ceramics have been studied [13] using an innovative equipment (GEMH Limoges France). As schemed in fig. 25, an alumina guide allows to separate the electronic equipment and the sample placed at high temperature in a furnace. A transducer is used to produce ultrasonic waves and the echoes from the material lead to the Young modulus calculation. Done under cycling temperature, this gives numerous information (fig. 26) on the thermomechanical properties and behavior of the material and their possible modifications during the cycling.

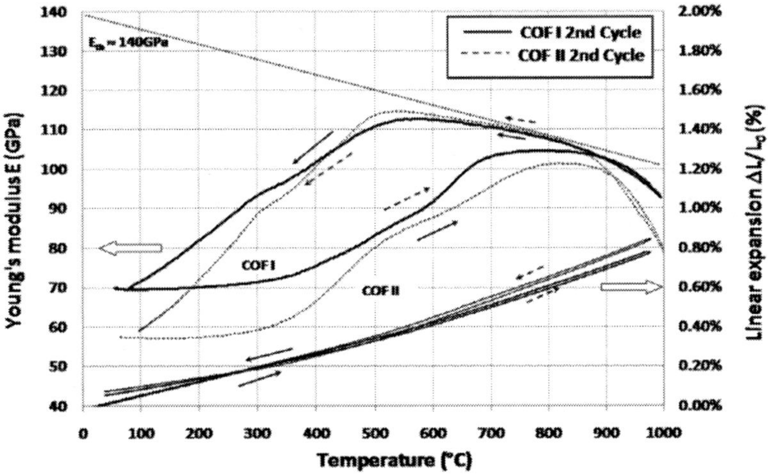

Fig. 26. – Young modulus and dilatometry of ACW ceramic under thermal cycling.

DOI: 10.1051/epjconf/201714800014

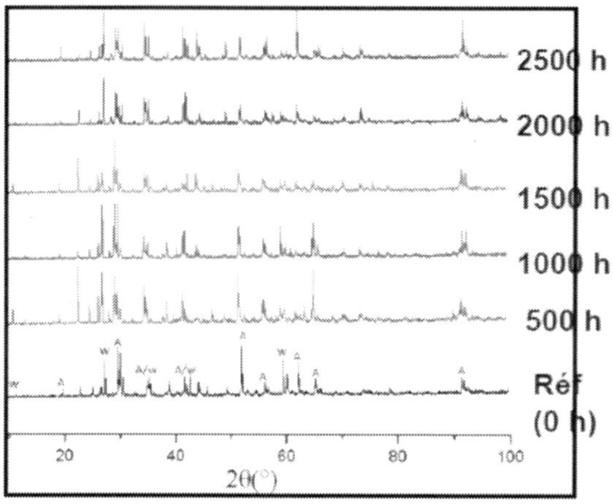

Fig. 27. – XRD diagrams of ACW ceramics submitted to high-pressure high-T cycling under hot air.

This characterization can be advantageously used to assess a material, to select the proper one for a particular applications and also to study its fatigue during its industrial operation.

Another important aspect, as mentioned above, is the compatibility of the TES material with the HTF and working conditions [14,15]. This is well illustrated by the evolution of the ACW ceramics submitted to high-temperature thermal cycling under pressurized air. Under atmospheric pressure, those ceramics present no change in structure but as shown in figs. 27 and 28, under pressure the ceramic initially made of 50% Augite and

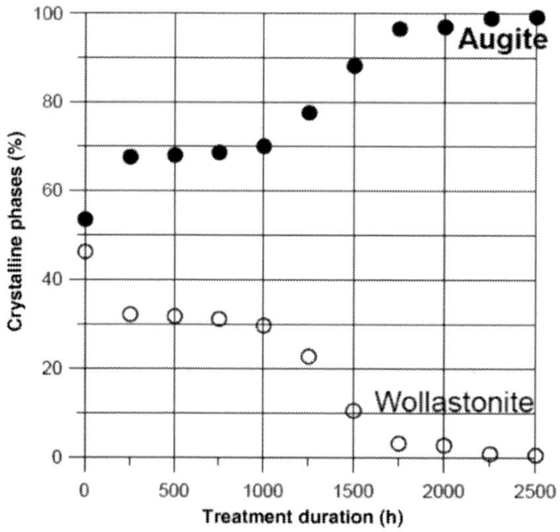

Fig. 28. – Evolution of the crystal structure of the ACW ceramics submitted to hot pressurized air.

Fig. 29. – Flat plate and corrugated plates of recycled ceramic.

Wollastonite is made after 2000 h of treatment of Augite only. This demonstrates how important could be to assess the ability of a storage material to be used under realistic working conditions.

Those materials have been produced under various shapes (fig. 29) and sizes such as spheres, cylinders, plates, corrugated plates, etc. This allows to develop structured heat storage packing (fig. 30) for which the plate thickness and geometry are designed to optimize simultaneously the heat transfer coefficient with the HTF, the pressure drop and the charge/discharge of the material deep to its center. The advantage of those geometries is also due to the fact that they are self-supported (avoiding any mechanical stress on the TES envelop) and they are highly modular (allowing easy scale-up).

Fig. 30. – Structured packing of TES static mixer modules made of recycled ceramics.

The LCA of this approach has been studied and compared to the current two-tank molten salt technology [16]. It highlights that even if the environmental and energy pay-backs of the waste heat treatment needed for dangerous wastes is less than 2 years, the impact of this waste treatment on the comparative LCA is favorable for the molten salt. In the case of fly ash or slags, the advantage is clearly for the recycled ceramics.

Up to date, the ACW heat treated in Europlasma/Inertam-France led to ground inert basalt-like materials for which the sole market of road base ground particles was identified. As the owner of the initial dangerous waste has to pay for the treatment itself, the price at which the granular material is sold is about 10 €/t. Compared to usual prices of high-temperature refractory ceramics ranging from 2000 to 9000 €/t, this demonstrates the potential of producing well-shaped proper recycled ceramics for TES applications.

5·3. *Other approaches.* – Among other developed alternative, we can mention the TES technology using high-temperature concrete of the DLR (Germany) [17]. This approach consider advantageously that concrete is a very common material, highly available, at low cost, it can be easily shaped and it presents storage capacities similar to other ceramics or rocks.

Nevertheless, usual concrete used for buildings is not supposed to be used an d cycled at high temperature. Moreover, concrete contains a high amount of water and a significant part of it should be removed properly during the first heating. This imposes a time- and energy-consuming first treatment during which the energy corresponding to the latent heat of evaporation has to be paid. In another field, concrete presents rather low thermal conductivity (about 1 W/(m K)) and then, an extended multitubular heat exchanger has to be embodied into the concrete blocks to guarantee the needed heat transfer rate. Obviously, this embodied heat exchanger increases the cost of the storage and the complexity of the dismantling step of the unit. This concrete-based TES for CSP is currently tested at pilot scale in Almeria (Spain) associated to nitrate phase change storage materials for the particular case of direct steam generation (DSG) future CSP plants.

6. – Conclusions and perspectives

The thermal energy storage for CSP offers new opportunities for innovations which are critically needed for their industrial growth to achieve the expected energy transition.

Several emerging approaches are already under progress and will be also applied to other kinds of applications such as for ACAES (adiabatic compressed-air electricity storage) [18] or the valorization of industrial waste heat [19].

While some of those new TES approaches benefit from all the available knowledge (this is the case of recycled ceramic), others, such as nano-fluids need major efforts in fundamental research to understand and control the involved phenomena properly before applying them to industrial applications.

REFERENCES

[1] Py X., Azoumah Y. and Olives R., "Concentrated Solar Power: current technologies, major innovative issues and applicability to West African Countries", *Renew. Sustain. Energy Rev.*, **18** (2013) 306.

[2] Gil A., Medrano M., Martorell I., Lazaro A., Dolado P., Zalba B. and Cabeza L. F., "State of the art on high temperature thermal energy storage for power generation. Part 1 - concepts, materials and modellization", *Renew. Sustain. Energy Rev.*, **14** (2010) 31.

[3] Calvet N., Gomez J., Faik A., Roddatis V. V., Meffre A., Glatzmaier G. G., Doppiu S. and Py X., "Compatibility of a post-industrial ceramic with nitrate molten salts, for use as filler materials in a thermocline storage system", *Appl. Energy*, **109** (2013) 387.

[4] Burkhardt J. J., Heath G. and Turchi C. S., "Life cycle assessment of a parabolic trough concentrating solar power plant and the impacts of key design alternatives", *Environ. Sci. Technol.*, **45** (2011) 2457.

[5] IEA International Energy Agency, *Technology Roadmap: Concentrating Solar Power* (OECD/IEA) 2010, `www.iea.org/roadmaps`.

[6] Pihl E., Kushnir D., Sanden B. and Johnsson F., "Material constraints for concentrating solar thermal power", *Energy*, **44** (2012) 944.

[7] Arthur O. and Karim M. A., "An investigation into the thermophysical and rheological properties of nanofluids for solar thermal applications", *Renew. Sustain. Energy Rev.*, **55** (2016) 739.

[8] Py X., Calvet N., Olives R., Meffre A., Echegut P., Bessada C., Veron E. and Ory S., "Recycled Material for Sensible Heat Based Thermal Energy Storage to be Used in Concentrated Solar Thermal Power Plants", *J. Sol. Energy Eng.*, **133** (2011) 1.

[9] Kere A., Py X., Olivès R., Goetz V., Sadiki N. and Mercier-Allart E., "High temperature thermal energy storage material from vitrified coal-fired power plant Fly-Ash", *Innostock 2012, 12th International Conference on Energy Storage, 16–18 May 2012, Lleida, Spain.*

[10] Meffre A., Py X., Olives R., Bessada C., Veron E. and Echegut P., "High Temperature Sensible Heat Base Thermal Energy Storage Materials Made of Vitrified MSWI Fly Ashes", *Int. J. Waste Biomass Valoriz.*, **6** (2015) 1003.

[11] Kere A., Dejean G., Sadiki N., Olives R., Goetz V., Py X. and Mercier-Allart E., "Vitrified industrial wastes as thermal energy storage materials for high temperature applications", *WasteEng 2012, 4th International Conference on Engineering for Waste and Biomass Valorisation, 10–13 September 2012, Porto, Portugal.*

[12] Gutierrez A., Miró L., Gil A., Rodríguez-Aseguinolaza J., Barreneche C., Calvet N., Py X., Fernández A. I., Grágeda M., Ushak S. and Cabeza L. F., "Advances in the valorization of waste and by-product materials as thermal energy storage (TES) materials", *Renew. Sustain. Energy Rev.*, **59** (2016) 763.

[13] Meffre A., Tessier-Doyen N., Py X., Huger M. and Calvet N., "Thermomechanical characterization of waste based TESM and assessment of their resistance to thermal cycling up to 1000 °C", *Waste Biomass Valoriz.*, **7** (2016) 9.

[14] Motte F., Falcoz Q., Veron E. and Py X., "Compatibility tests between Solar Salt and thermal storage ceramics from inorganic industrial wastes", *Appl. Energy*, **155** (2015) 14.

[15] Kere A., Sadiki N., Py X. and Goetz V., "Applicability of thermal energy storage recycled ceramics to high temperature and compressed air operating conditions", *Energy Convers. Manag.*, **88** (2014) 113.

[16] Lalau Y., Py X., Meffre A. and Olives R., "Comparative LCA between current and alternative waste-based TES for CSP", *Waste Biomass Valoriz.*, **7** (2016) 1509.

[17] LAING D., STEINMANN W. D., TAMME R. and RISHTER C, "Solid Media thermal storage for parabolic trough power plants", *Sol. Energy*, **80** (2006) 1283.
[18] JUBEH N. M. and NAJJAR Y. S. H., "Green solution for power generation by adoption of adiabatic CAES system", *Appl. Therm. Eng.*, **44** (2012) 85.
[19] MIRO L., GASIA J. and CABEZA L. F., "Thermal energy storage (TES) for industrial waste heat (IWH) recovery: a review", *Appl. Energy*, **179** (2016) 284.

Chemical energy storage: Part of a systemic solution

ROBERT SCHLÖGL

*Fritz-Haber-Institut der Max-Planck-Gesellschaft - Faradayweg 4-6,
D-14195 Berlin, Germany and
MPI für Chemische Energiekonversion - Stiftstrasse 34-36,
D-45470 Mülheim an der Ruhr, Germany*

Summary. — This paper is a primer into concepts and opportunities of chemical energy storage. Starting from the quest for decarbonisation we reveal the possibilities of chemical energy storage. We briefly discuss the critical role of catalysis as enabling technology. We concentrate on options of large-scale production of chemicals from CO_2 and green hydrogen. We discuss one potential application of fueling future combustion engines that could run with minimal regulated emissions without exhaust purifications and legal tricks.

1. – Introduction

It is now clear that the world will act against climate change created by excessive combustion of fossil energy carriers. The world emits ca. 1×10^{10} t carbon atoms [1] per year from which ca. 75% stem from burning fossil energy carriers. Figure 1 shows a time series over recent history. Here the emitted carbon mass is given as CO_2: the conversion factor to the mass of carbon is 3.66.

About half of the 75% come from power generation, the rest is shared between mobility and domestic/industrial heat production. Local and temporal variations are significant, creating a multitude of emission scenarios in the world. Here it is critical to see that replacing fossil power stations by solar electricity from wind, PV and other sources is not

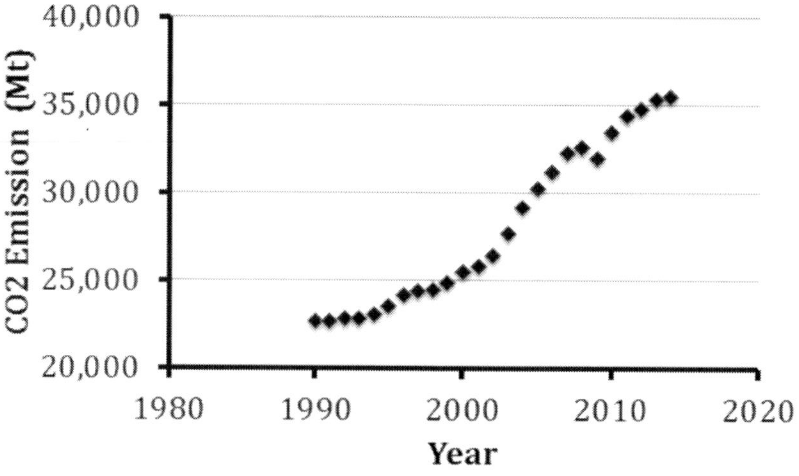

Fig. 1. – Temporal evolution of the global CO_2 emission. All emission sources are collected in this graph.

sufficient to solve the climate change problem. Another key issue is the volatility of the green electricity being at variance to the demand profiles of power in many parts of the world. We may estimate that we can balance out without major energy storage about 50% of the load in a typical grid that is powered by green electricity [2]. The other 50% needs to be controlled power generation by combustion or nuclear energy or we need to overpower the solar generation capacity enormously and then we may end at about 75% green electricity. It is a key infrastructure to western societies (at least) to have electrical power uninterrupted whenever needed. The majority of the power goes into industrial applications where many processes run continuously 24/7 and cannot easily adapt to significant a contribution of fluctuating energy supply.

If we demand that this property as well as unlimited mobility are conditional boundaries to sustainable energy systems then we have to accept that we need energy storage at world scale and not only locally for small-scale domestic applications in residential and rural areas. This is discussed [3] for example with combinations of PV and local battery systems. Such concepts are perfectly useful and understandable but create some larger-scale problems. One is the societal dimension that people tend to forget the systemic nature of sustainable energy and think they have done their share. Economically such systems put the burden of energy system transformation on those who cannot build or afford such systems. Technically they sharpen the challenge for utility providers to hold capacities in place that are very rarely needed and thus maximal uneconomical. Useful would be smart systems of such architecture where the utility provider can control storage and supply according to the needs of the grid and not to the local needs.

This excursion highlights the complexity of measures in a systemic environment where locally "good" measures may have multiple detrimental effects in some other parts of the system. Such arguments also hold for the quest of decarbonisation through electrification

DOI: 10.1051/epjconf/201714800015

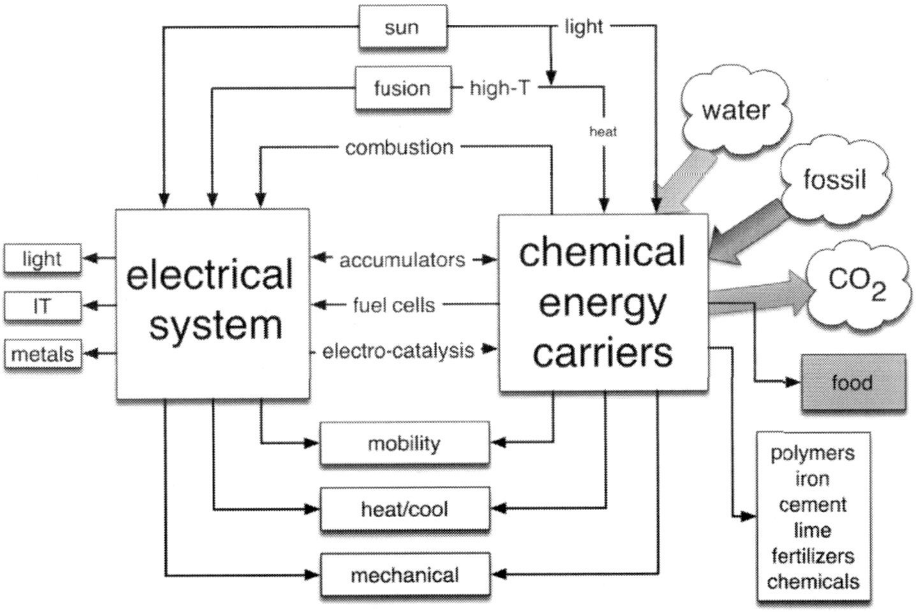

Fig. 2. – General structure of energy systems. Core applications are given. Either energy carrier type (large squares) can power some of them, some others are uniquely tied to the other carrier. Solar energy occurs both in form of photons and as heat resulting from interaction of solar photons with the atmosphere and hydrosphere of the planet.

of our society. .Figure 2 shows that every energy system is always a combination of free electrons and chemical bonds as energy carriers. We are not free to omit one of them, as there are specific unique applications to each carrier type. We further see that we are good in converting chemical bonds into free electrons but that we cannot do the reverse very well. This creates the fundamental challenge of chemical energy conversion. We need to learn how to do conversion in both directions. This is independent of the not fruitful discussion about efficiencies; if we want to be sustainable then we have to minimize and in the long run abandon fossil energy carriers. True is that this process is not rapid, it will cover human generations and require multiple intermediate transformation paths. The size of the energy systems is simply too large for disruptive technical changes without disrupting the socio-economic conditions that are at least as critical for a successful transformation than technical and engineering challenges.

One reason why bi-directional conversion is so difficult is that converting chemical bonds into free electrons follows the laws of thermodynamics with fossil energy carriers. These laws state that chemical reactions are facile if they are downhill in free energy once we provide sufficient activation. Combustion is the perfect example for that, as CO_2 and water are thermodynamically stable products. Back-conversion is more demanding as then we wish to store energy in chemical bonds, which by definition is energetically uphill and does not occur spontaneously. We see below how that is accomplished.

DOI: 10.1051/epjconf/201714800015

Here we note that all fossil energy followed this uphill energy storage process that was enabled by solar energy and biomass in pre-historic times. Consequently, we should critically abstain to discuss "efficiency" of chemical energy storage in comparison to existing fossil energy carriers. This is often done to discredit sustainable energy systems in economic and technical terms being however a scientifically invalid argument.

One strategic element of chemical energy conversion is interfacial catalysis, regardless if it occurs in batteries, electrolyzes, photochemical devices or solar refineries. All chemical processes are subject to the dual delimiters of thermodynamics (uphill, downhill) and kinetics (slow, fast) that must not be intermixed. We cannot change thermodynamics but we can change kinetics by catalysis.

The science of catalysis [4] deals with the phenomenon that we can influence the rate of a chemical reaction by changing its course from a modus in which the reactants meet and either react or separate without reaction (homogeneous reaction) to a situation in which we use an auxiliary material (catalyst) that can be a molecule, a cluster, an enzyme, or a solid surface. This auxiliary material binds the reactants to it and allows for meeting of the reactants in their bound states. In many cases it acts not only in this "marriage broker" way but it also is a "bank" for redox equivalents (electrons) and donates to reactants electrons that the product has to give back to the bank in order to be released from the catalyst. The trick to controlling chemical reactions by catalysis is to separate meeting and reacting of the reactants in space and time and to "prepare" the reactants for their intended chemical transformation.

The challenge is to perform the reaction such that the catalyst is regenerated and can be used for more than one set of reactant transformation. In other words the binding between catalyst and reactant must be "intermediate" meaning strong enough to prepare the reactants but weak enough to allow their removal after reaction. The resulting "turnover number" per unit active site on a catalyst can be anything from a few hundreds to a few millions. We cannot well measure this important number as we have no clear picture about the active site. This is the structure of the catalyst that does the binding and exchange of reactants. For molecules as catalysts the number of options is naturally much less than for solids where a literally infinite number of configurations can exist. But as we cannot observe active sites at work and we do not know if they are continuously working or they take time off and stay in a "resting state" we have great problems to define the quantity "activity" in any catalyst.

Where we have fewer problems is to understand their energetic function. Let us assume that we want to store energy in CO_2 by forming methanol as platform molecule using green hydrogen that we generate from water splitting. Figure 3 indicates a coarse profile through the potential energy landscape of this reaction. The details are known [5] but too complex to be used here.

The reaction is

$$CO_2 + 3\ H_2 \longleftrightarrow CH_3OH + H_2O.$$

It is limited by thermodynamic equilibrium and does not run to completion under

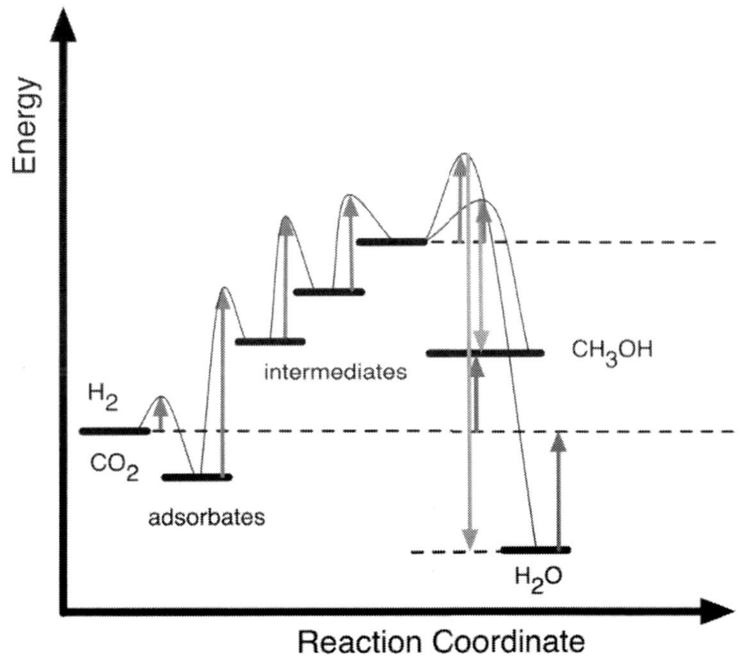

Fig. 3. – Schematic section through the potential energy landscape of the reduction of CO_2 with hydrogen to methanol. We may use a modified copper catalyst, or Ru-organic molecules to perform the reaction. The red arrows denote kinetic energy barriers, the green arrows kinetic energy gain and the blue arrows thermodynamic energy quantities.

practical conditions. The reason is that we lack a catalyst that performs the reaction fast enough at low temperatures (around 400 K). We need 600 K and a pressure of 60 bar typically to obtain a technical relevant productivity. Another complication is that this reaction has a second product namely CO that is undesired. This further limits the conditions of operation of methanol synthesis.

From the diagram we see that we store energy in methanol, as it is higher in energy by about 0.5 eV than the mix of CO_2 and hydrogen. This is an uphill reaction that cannot occur for reasons of thermodynamics. The trick is to couple the reaction with a process that is downhill in energy and thus provides the driving force for methanol synthesis. This is the formation of water, which is not an unwanted byproduct but the reason for the ability to store energy.

The blue arrows in fig. 3 say nothing about the velocity of the reaction. The red arrows designate the activation energies from moving through the reaction coordinate (perform the making and breaking of chemical bonds) and say something about the reaction velocity Much of this barrier energy is re-gained upon completion of the energy (green arrows). The rest of it is energy loss, as we have to heat and pressurize the system to reach useful reaction velocities (about 1 kg of methanol per kg catalyst and hour). We see that we need to climb a high barrier of several intermediates before we reach at the climax an almost ready methanol molecule and a deactivated catalyst that

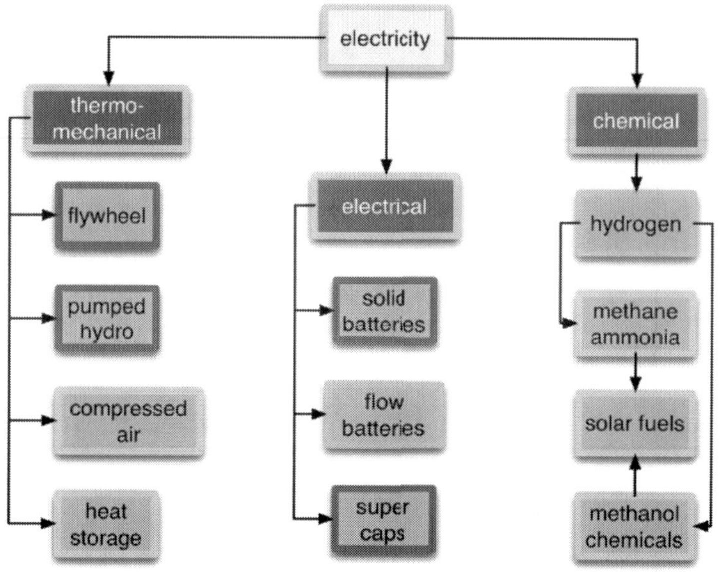

Fig. 4. – Energy storage options with physical and chemical means. The red boxes denote solutions that are used in present energy systems, the light blue ones are options almost ready for grid scale use, the other options are in demonstration or research status.

is charged with OH groups. It is the ingenuity of the chemist to minimize the number of intermediates and related activation barriers and to avoid that other reaction products occur during the journey over the higher energy barriers. When these barriers get very high it is unlikely that we obtain the wanted product and we produce unwanted products, such as carbon in the form of soot. This would be an expensive way of immobilizing CO_2 that could be stored without danger in solid form, but as we will see, there are more effective ways of doing so.

If we now return to our energy issue we can ask what means of energy storage we have and if there are more convenient ways than chemical energy storage with all complexity standing behind the facts from fig. 3. In fig. 4 we give a systematic overview about storage options.

The multitude of options [6] becomes rapidly smaller when we consider their characteristics being storage capacity and storage times. We discriminate options for shaping the power system (very short time), for stabilizing the power system (hour timescale) and for long-term storage (seasonal time scales). All chemical energy storage solutions belong to the latter, as does pumped hydroelectricity. The other options are mostly suited for the other short-time purposes.

In fig. 5 we see a generic design of a sustainable energy system with some key components. Such a design takes care about the need of marginating both chemical bonds and free electrons as energy carriers. It is realistic in terms of utilizing solar energy as much as possible and allows for offering energy in the quality and distribution in space and time that we are used with fossil systems.

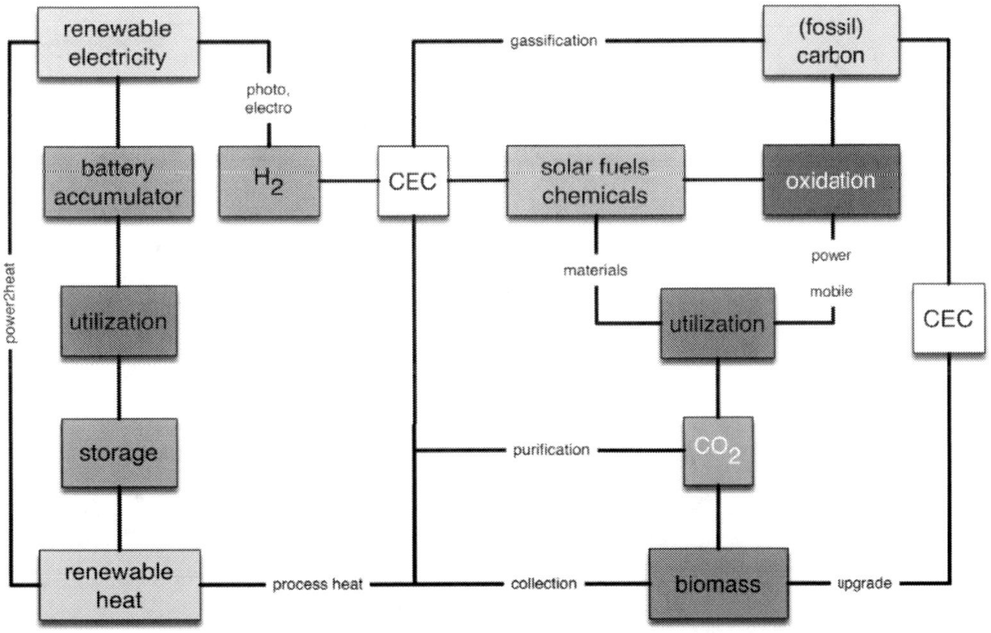

Fig. 5. – A generic design of a sustainable energy system. It uses primarily solar energy and converts it into free electrons and several forms of chemical bonds as energy carriers. Hydrogen is a critical intermediate but no consumer product. It uses as much electricity as possible and serves the needs of those elements in the energy system that cannot be electrified. It uses biomass as carbon collector and not as energy source. CEC stands for chemical energy conversion.

We will now concentrate on the long-term storage options as multiple information is available on battery science [7] being a critical part not only for grid stabilization but also for bringing green energy to the mobility sector. From fig. 5 we see that we can use green energy and green heat from the sun. We should use as much green electricity as possible immediately upon generation and provide a suitable large-area grid system with the necessary intelligent control system. Solar heat in low and high temperature form is easier to store than electrons for use as domestic heating (and cooling) application but also in some parts of the world as backup power station. All the rest of solar energy should be converted [8] in chemical bonds through the formation of hydrogen gas. Besides electrolysis [9] there is also plasma chemistry and combined photo-electrical conversion processes [10] at low and high temperatures that can be used. None of these concepts are ready yet for grid scale use. Also electrical water splitting is a difficult technology under the conditions of large-scale use [11] and variable electricity input; existing large systems run only on absolutely continuous electricity supply from hydroelectric installations.

Water splitting is a redox reaction in which oxide from water is oxidized to oxygen gas and the electrons gained from this are used to discharge protons into hydrogen gas:

$$2\ H_2O \longrightarrow O_2 + 4\ H^+ + 4e^-,$$
$$4\ H^+ + 4e^- \longrightarrow 2\ H_2.$$

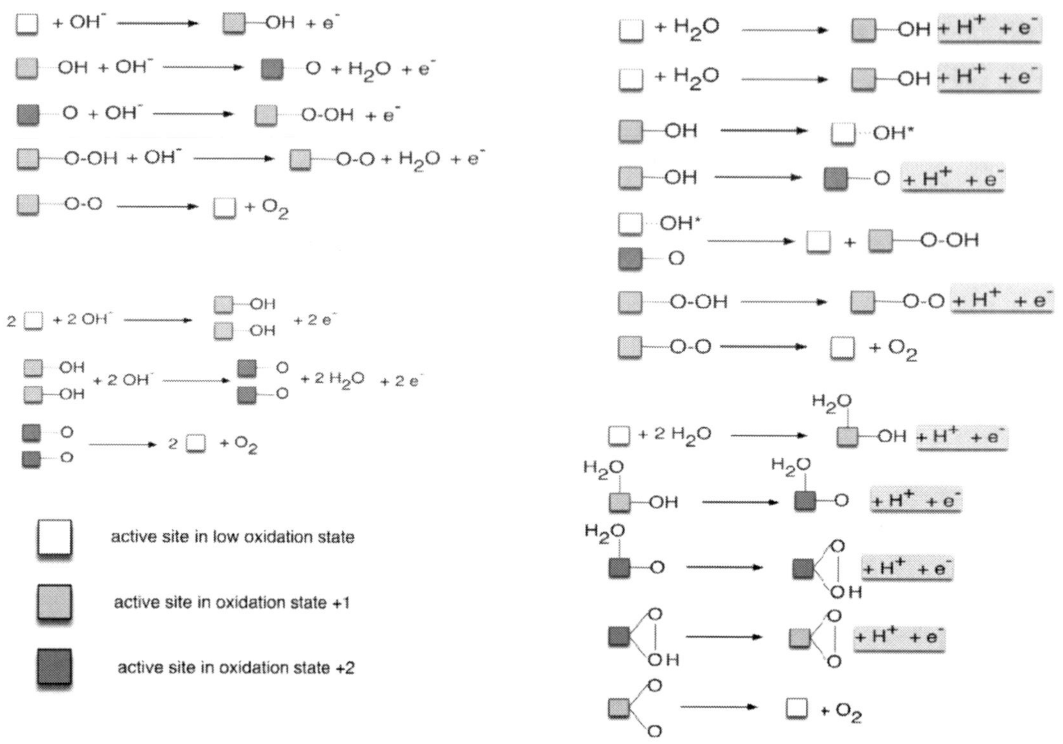

Fig. 6. – Reaction mechanisms of oxygen evolution in water splitting. The active site is denoted with colored squares, occurring in different oxidation states as indicated. The processes on the left side occur in basic media, the processes on the right side in acidic media.

The reaction is separated into two half reactions in electrical devices requiring two reaction spaces and two electrodes where the charges are separated from the ions. An advantage of this device is that the two reaction products hydrogen and oxygen are separated from each other which is highly advantageous. The disadvantage is that a membrane is necessary between the two compartments in order to allow the ions to migrate between the electrodes. The material science of these membranes and the limits they impose on reaction conditions are critical for the optimization of the device.

This reaction is carried out in nature as well as in electrolyzes or in photo-electrochemical cells. The exact nature [12] of the separation of charges and ions is different in these devices. We have learned much about the way in which nature does this reaction during photosynthesis [13] and we can use this knowledge for designing technical devices. We have to abstain from the idea to copy the biological process as it is way too complex for technical realization and it is also not efficient enough on the basis of hydrogen production per catalyst atom: the used enzymes are extremely complex and require continuous self-repair to remain operational.

We are far away from understanding on a robust basis the challenge of electrochemical water splitting with technical electrodes. We have a good idea about the possible

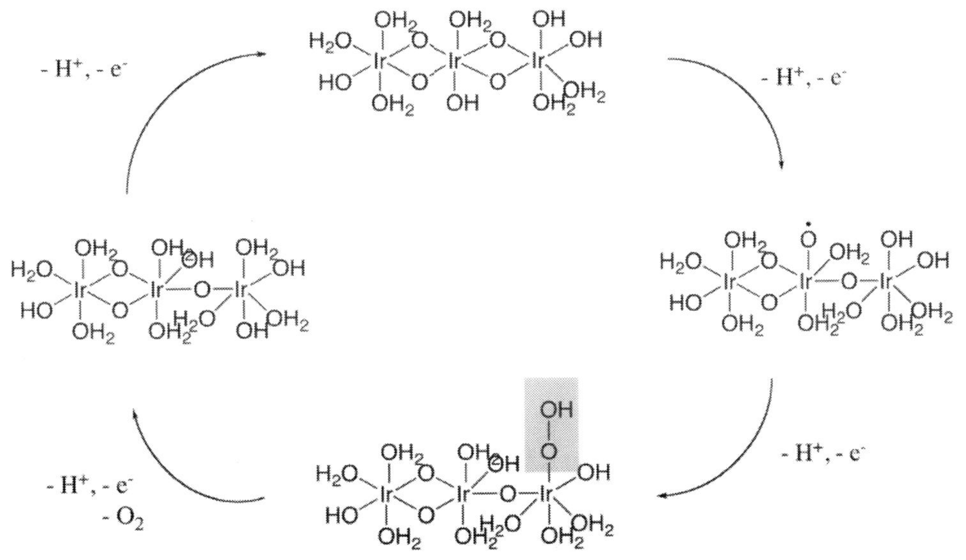

Fig. 7. – Likely reaction mechanism of oxygen evolution for water splitting. The critical intermediate is indicated.

mechanisms [14] and we know that not the hydrogen-forming part but the oxygen-forming part is the difficult reaction. We also have evidence that the reaction of oxygen formation is so difficult, as it requires an intermediate (an oxygen molecule with only one bond between the atoms) that is a very reactive species and that is likely to react with the electrode or hydrolyze with water rather than to finish the oxidation process and give off the final electrons.

In the last decades all likely mechanisms for oxidizing water in the so-called "oxygen evolution reaction" (OER) were formulated. It is surprising that such a simple reaction can be so complicated in detail. Figure 6 collects the likely reaction pathways. They are grouped according to basic or acidic environments. One can see that the choice of the condition of the water reactant plays a crucial role in the course of the reaction. Today one thinks that more than one reaction mechanism can co-exist in a working system. What is also clear from the figure is that the active site must be capable of changing its oxidation state, here assumed for two units per site. This is no problem for a metallic catalyst having the large number of electrons available at the Fermi level. For oxides that are required to withstand the aggressive character of the intermediates and allowing for facile adsorption of water this is a complicating requirement that can be fulfilled by several elements in the periodic table notably by the oxides of the late transition metals.

In fig. 7 we show a likely reaction cycle of oxygen evolution using the best known catalyst being $IrO(OH)(H_2O)_n$. The critical reaction intermediate that can either form oxygen or form a peroxide species with the catalyst is shown. We note that the structure of the active species is quite flexible and does not resemble the notion of an inorganic rigid oxide. It can change the coordination of the metal atoms during reaction. It is believed that both properties may explain the excellent performance of the catalyst.

Such hydrated species with other metal atoms tend not to be stable and condense into oxides under evolution of water. Such reactions are well known in the corrosion of iron to rust, for example. Many materials are being developed at present [15] with the hope to overcome the stability problems and poor performance of cheaper systems than Ir belonging to the rarest elements on the planet.

Let us assume we have solved the hydrogen problem and we have access to green hydrogen in the necessary amounts. Then we may form methanol from CO and hydrogen as platform molecule for many downstream products. In order to get a feeling for the quantities involved in such processes we discuss briefly the requirements of materials and the cost structure of such a process that does not yet exist in reality on grid scale. The world produces about 6×10^7 t/a methanol from CO_2 and CO with hydrogen coming exclusively from fossil sources mainly from reforming methane with water. This process is cheap and easy to conduct in large installations and thus methanol plants can be very large factories with typically 1×10^6 t/a output. When we think about a methanol plant with green hydrogen the size of the electrolyzer plant may be controlling. Existing large electrolyzers have for example between 10 and 15 MW power consumption that could come from a wind park. Then we see the following gross numbers for our plant. These are realistic numbers as not thermodynamic values but practically measured performances are used for the estimation. Nonetheless there will be additional losses for the production environment and the consumption of the plant utilities that are not taken into account. We obtain:

Electrolyzer 2100 Nm^3/h: 13 MW/h power.

Produces 950 kMol H_2/h.

Can reduce 315 kMol/h CO_2 to methanol.

Equals to 7.5 MMol/d or a usage of 330 tpd.

Equals to 120.000 tpa CO_2 usage. (ca. 15 M€ p.a.)

This produces 88.000 tpa MeOH. (ca. 32 M€ p.a.)

Power amounts to 115.000 MWh. (ca. 5 M€ p.a.)

Even when we consider that the cost of electricity assumed to be 4 ct/kWh may be too low, we recognize that the economics seems viable. There is, however, one critical condition. If many of such plants would operate, then the prize for methanol would fall drastically if no new utilization for methanol can be found. Here we see a general problem of chemical energy storage. The idea to replace existing chemical raw materials by green molecules finds quickly its end when we discuss chemical energy storage as a means to utilize CO_2 (CCU in power2X processes) and not as a method to transfer green energy in other sectors of the energy system. The whole chemical industry in the world consumes about 4% of fossil resources and thus can only take up the same amount of carbon from CO_2. On the other hand the generation of chemicals from fossil sources produces a lot of additional CO_2 for using fossil sources (mainly gas) to provide the necessary heat and hydrogen through steam reforming. One way of using green hydrogen would thus be to power the chemical industry and to replace black hydrogen for hydrogenation of carbon and nitrogen (in the Haber-Bosch process for ammonia synthesis) by green hydrogen. This alone would reduce the carbon footprint more than the replacement of the fossil material sources (mainly naphta, a fraction of oil distillation).

So even if we can partly decarbonize chemical industry (this is still a formidable task to do) we need additional concepts to move green energy into the supply system. When we inspect figs. 3 and 5 we may realize that shortening the transformation chain of chemical energy conversion is a good idea to stay energy-efficient. Then we should discard the idea of CO_2 hydrogenation and stick to hydrogen. The hydrogen economy [16] is a great idea and will likely play a role in upstream energy storage. The author is not convinced that this also holds for a end-user hydrogen economy as we have existing alternatives (see below) and do not want to bother end users with the unpleasant experience to deal with the infrastructure for the energy carrier pure hydrogen.

Using green electricity to generate hydrogen and back-convert this to electricity using a pipeline grid and (underground) storage is an option to counteract the volatility of green electricity. All technical elements of such a solution do exist except the water splitting devices compatible with stability under volatile load and with a cost structure acceptable for such an application. Such an option is, however likely more expensive that using cheap green electricity for heating applications where also CO_2 can be saved and where no technological barriers exist. In later stages of the energy system transformation we will likely see not a competition but a co-existence of both options. Then we have a critical need to be able to split water for green hydrogen production.

What needs to be mentioned here is that the current debate about "excess green electricity" (that does not physically exist in a working power grid) is not the correct driver for creating a hydrogen generation technology of world scale. If we want to supply the whole energy system comprising power, mobility, heating and materials with green energy then we need much more primary green electricity than can ever result from the volatility excess of the power generation branch in the energy system. We need green electricity for the sake of doing chemical energy conversion and not consider this family of technologies as a remedy for excess electricity.

When we consider the priorities of useful applications of green hydrogen then we need to specify what is our priority of targets in changing the energy system. Avoiding CO_2 emission as prime target clearly calls for re-generation of electricity, as all other options create CO_2, and coal, the most economic source for electricity, the largest specific amounts of all energy carriers (ca. 0.8 kg CO_2 per kWh). If we wish to replace CO_2 most economically and generate the co-advantage of becoming less dependent on energy imports then we need to funnel the green hydrogen into mobility. This would reduce the dependence on oil being the largest volume of imported energy carriers in most countries. As oil carries substantial hydrogen with it (as octane C_8H_{18} the combustion generates energy for 9 moles of water and 8 moles of CO_2) its use generates specifically less CO_2 than the use of coal in power generation. As we value mobility and thus fuel specifically higher than electricity the revenue of making artificial fuels is higher than re-generating electricity.

 DOI: 10.1051/epjconf/201714800015

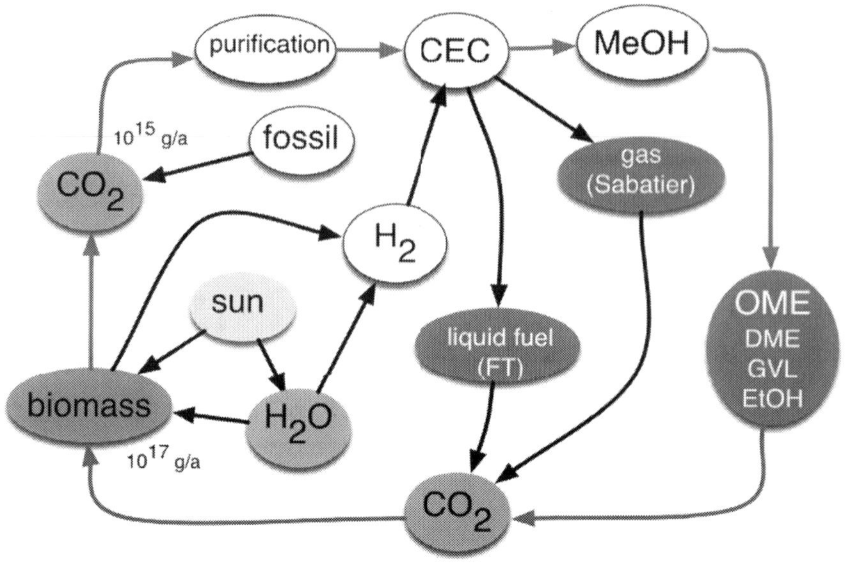

Fig. 8. – A closed carbon cycle for mobility. The blue arrows combine the elements forming the cycle. The red elements denote synthetic and designer fuels. The numbers specify the annual growth of biomass from which about 1% would be sufficient to collect all CO_2 from present global mobility. The extent to which we use still fossil sources for other applications in the energy system defies the deviation from a fully closed cycle. This option indicates a transition path to arrive at a fully sustainable system.

When we have CO_2 and hydrogen and want to create valuable liquid fuels then we have two choices. We can convert CO_2 to CO and use the synthesis gas with the family of Fischer-Tropsch technologies for making artificial fuels replacing gasoline, jet fuel or diesel. This is practiced today at world scale using fossil gas sources to prepare synthesis gas and comes under the name GTL process (gas-to-liqid). The resulting fuels are of very high quality and free of impurities like sulfur and pay extra value at the filling station. This approach may be augmented by the addition of biomass to the reactant mix helping to minimize the atom economical waste of green hydrogen in forming the necessary but unwanted water molecules.

More innovative is to redesign the whole power train in vehicles in order to maximize the use of green electricity. We can combine an electric car with a mid-sized battery for short term driving and for recuperating deceleration energy with devices using on board green fuels. Some prefer green hydrogen and a fuel cell with the advantage of zero emission at the car and the disadvantage of having to install a new infrastructure (quasi worldwide). The author suggests to use liquid designer furls that can be distributed in the present fuel system but that have some advantages over present fuels without being incompatible with them. We may design molecules that burn without regulated emissions of NO_x and carbon particles. This indeed possible [17] when we use the insights from mechanistic studies [18] of combustion science. A co-condensation product of methanol with formaldehyde known as the family of oxymethylene Ethers (OME) is a perfect surrogate [19] for Diesel with the only disadvantage of having ca. half the energy density than Diesel. Similar solutions of molecules with C-O-C moieties rather than C-C-C groups can be found replacing gasoline. Only for jet fuel we have to resort to carbon-rich molecules for reasons of maximum energy density.

It we now complete our journey into chemical energy conversion then we can see how this technology can solve our energy problem without the quest for decarbonisation that is only partly possible as we have learned with fig. 2. We should reduce the unbalanced use of fossil energy as much as possible. But we need not to stop using the great advantages of carbon-based energy carriers. All we need to do is to follow the example of nature and close the carbon flux cycle. When we combine the use of carbon energy carriers with the collection of CO_2 and its re-reduction with green hydrogen then we can use without any harm carbon fuels in amounts limited by the availability of green hydrogen.

This is a far-reaching target but has the inherent advantage of not having to rely on reduction of energy consumption, which is adverse to the trends in global population and the socio-economic pressure of all humans having the equal right of energy access. If we wish to consider such architecture for sustainable energy systems [20] then we see that we are still at the very beginning of this transformation. We should concentrate to improve the bulk productivity of solar energy converters (PV, wind) as much as possible and not be satisfied with present technologies. All conversion technologies need enormous and disruptive changes in performance and cost structure for such a scenario. In fig. 8 we show such a scenario for the limited range of mobility in the sustainable energy system.

The simplicity of the scheme should not mask the enormous challenges for science and technologies as well as for economics and social comprehension and acceptance. This is a vision with little chance for realization in this generation. But without a master plan of priorities in science and technology we would not know how to evolve our capabilities in the energy challenge. As much as many fundamental discoveries from curiosity-driven science are needed as much also application-oriented science is required to select and evolve the science and understanding as basis for real technology evolution. At first we need some understanding about the scale and complexity of an endeavor "energy system transformation" that we seemingly stumble into without knowing much bout the inner structure of it. As little time as the climate change leaves us for the energy transformation, as detrimental would it be not to first structure the challenge and agreeing on a list of priority targets that will be different in different parts of the world. Global collaborations between scientists is the first and foremost requirement to ever meet the challenge and this in time for limiting the effects of climate change.

REFERENCES

[1] Aresta M. and Dibenedetto A., *Dalton Trans.*, **2007** (2007) 2975, DOI:10.1039/B700658F; Riahi K., Kriegler E., Johnson N., Bertram C., den Elzen M., Eom J., Schaeffer M., Edmonds J., Isaac M., Krey V., Longden T., Luderer G., Mejean A., McCollum D. L., Mima S., Turton H., van Vuuren D. P., Wada K., Bosetti V., Capros P., Criqui P., Hamdi-Cherif M., Kainuma

M. and Edenhofer O., *Technol. Forecast. Soc. Change*, **90** (2015) 8; Heimann M. and Reichstein M., *Nature*, **451** (2008) 289.

[2] Wagner F., *Eur. Phys. J. Plus*, **129** (2014) 20.

[3] Kaundinya D. P., Balachandra P. and Ravindranath N. H., *Renew. Sustain. Energy Rev.*, **13** (2009) 2041.

[4] Schlögl R., *Angew. Chem. Int. Ed.*, **54** (2015) 3465; Chorkendorff I. N. and J. W., *Concepts of Modern Catalysis and Kinetics* (Wiley-VCH, Weinheim) 2007.

[5] Studt F., Behrens M., Kunkes E. L., Thomas N., Zander S., Tarasov A., Schumann J., Frei E., Varley J. B., Abild-Pedersen F., Nørskov J. K. and Schlögl R., *ChemCatChem*, **7** (2015) 1105; Schumann J., Eichelbaum M., Lunkenbein T., Thomas N., Alvarez-Galvan M. C., Schlögl R. and Behrens M., *ACS Catal.*, **5** (2015) 3260; Lunkenbein T., Schumann J., Behrens M., Schlögl R. and Willinger M. G., *Angew. Chem. Int. Ed.*, **54** (2015) 4544.

[6] Gallo A. B., Simoes-Moreira J. R., Costa H. K. M., Santos M. M. and dos Santos E. M., *Renew. Sustain. Energy Rev.*, **65** (2016) 800.

[7] Dunn B., Kamath H. and Tarascon J. M., *Science*, **334** (2011) 928; Shaahid S. M. and El-Amin I., *Renew. Sustain. Energy Rev.*, **13** (2009) 625; Bruce P. G., Freunberger S. A., Hardwick L. J. and Tarascon J. M., *Nat. Mater.*, **11** (2012) 19; Tarascon J. M. and Armand M., *Nature*, **414** (2001) 359.

[8] Clarke R. E., Giddey S., Ciacchi F. T., Badwal S. P. S., Paul B. and Andrews J., *Int. J. Hydrogen Energy*, **34** (2009) 2531.

[9] Centi G., Quadrelli E. A. and Perathoner S., *Energy Environ. Sci.*, **6** (2013) 1711.

[10] Yella A., Lee H.-W., Tsao H. N., Yi C., Chandiran A. K., Nazeeruddin M. K., Diau E. W.-G., Yeh C.-Y., Zakeeruddin S. M. and Graetzel M., *Science*, **334** (2011) 629.

[11] Gahleitner G., *Int. J. Hydrogen Energy*, **38** (2013) 2039.

[12] Rossmeisl J., Dimitrievski K., Siegbahn P. and Norskov J. K., *J. Phys. Chem. C*, **111** (2007) 18821.

[13] Ogata H., Kraemer T., Wang H., Schilter D., Pelmenschikov V., van Gastel M., Neese F., Rauchfuss T. B., Gee L. B., Scott A. D., Yoda Y., Tanaka Y., Lubitz W. and Cramer S. P., *Nat. Commun.*, **6** (2015) 7890; Lubitz W., Reijerse E. J. and Messinger J., *Energy Environ. Sci.*, **1** (2008) 15.

[14] Viswanathan V., Hansen H. A., Rossmeisl J. and Norskov J. K., *ACS Catal.*, **2** (2012) 1654; Rossmeisl J., Qu Z. W., Zhu H., Kroes G. J. and Norskov J. K., *J. Electroanal. Chem.*, **607** (2007) 83.

[15] McCrory C. C. L., Jung S., Peters J. C. and Jaramillo T. F., *J. Am. Chem. Soc.*, **135** (2013) 16977.

[16] Muradov N. Z. and Veziroglu T. N., *Int. J. Hydrogen Energy*, **33** (2008) 6804; Barreto L., Makihira A. and Riahi K., *Int. J. Hydrogen Energy*, **28** (2003) 267.

[17] Haertl M., Seidenspinner P., Jacob E. and Wachtmeister G., *Fuel*, **153** (2015) 328.

[18] Sarathy S. M., Osswald P., Hansen N. and Kohse-Hoinghaus K., *Prog. Energy Combust. Sci.*, **44** (2014) 40.

[19] Iannuzzi S. E., Barro C., Boulouchos K. and Burger J., *Fuel*, **167** (2016) 49.

[20] Schlögl R., *EPJ Web of Conferences*, **98** (2015) 04001; *Angew. Chem. Int. Ed.*, **54** (2015) 4436.

AUTHOR INDEX

EDP Sciences – Web of Conferences
17, Avenue du Hoggar
Parc d'Activité de Courtabouf
BP 112
F-91944 Les Ulis Cedex A
France

ISBN 978-1-5108-4514-5

CURRAN ASSOCIATES INC.
proceedings.com